This document is intended as a guideline for employers to establish their own written practice for the qualification and certification of their nondestructive testing personnel. It is not intended to be used as a strict specification.

Recommended Practice

SNT-TC-1A

December 1988 Edition

American Society for Nondestructive Testing, Inc.

*Society for Nondestructive Testing (SNT) - Technical Council (TC) - First Document (1A)

Published by The American Society for
Nondestructive Testing, Inc., 4153 Arlingate
Plaza, Columbus, OH 43228.

ISBN 0–931403–50–2

Printed in the United States of America.

Foreword

The recommended practice contained herein establishes the general framework for a qualification and certification program. In addition, the document provides recommended educational, experience, and training requirements for the different test methods. Supplementary documents include question and answer lists which may be used in composing examinations for nondestructive testing personnel.

Inquiries related to this recommended practice should be directed to the Chairman of the Personnel Qualification Division at the following address:

The American Society for Nondestructive Testing
4153 Arlingate Plaza, Caller #28518
Columbus, Ohio 43228-0518

Review Committee

Publication and review of this Recommended Practice was under the direction of the SNT-TC-1A Review Committee of the Personnel Qualification Division of The American Society for Nondestructive Testing, Inc.:

Henry M. Stephens, Jr., Chairman
Rick E. Cameron
Louis J. Elliott
O. Lee Gaston, Jr.
Edward A. Macejak
Phillip A. Oikle
Ramasamy Palanisamy
Adrian A. Pollock
Donald J. Quirk
Allen D. Reynolds
Wade J. Richards
Nickolas Rohach, Jr.
Carl B. Shaw
Jack C. Spanner, Sr.
Michael L. Turnbow
John H. Weiler

ASNT Staff Contributor:
Ronald H. Selner, Technical Services Manager

Contents

Personnel Qualification and Certification in Nondestructive Testing

Recommended Training Courses

Radiographic Testing Method

Appendix

Personnel Qualification and Certification in Nondestructive Testing

Recommended Practice No. SNT-TC-1A

Personnel Qualification and Certification in Nondestructive Testing

1. Scope

1.1 It is recognized that the effectiveness of nondestructive testing (NDT) applications depends upon the capabilities of the personnel who are responsible for, and perform, NDT. This Recommended Practice has been prepared to establish guidelines for the qualification and certification of NDT personnel whose specific jobs require appropriate knowledge of the technical principles underlying the nondestructive tests they perform, witness, monitor, or evaluate.

1.2 This document provides guidelines for the establishment of a qualification and certification program.

1.3 These guidelines have been developed by The American Society for Nondestructive Testing, Inc., to aid employers in recognizing the essential factors to be considered in qualifying personnel engaged in any of the NDT methods listed in Section 3.

1.4 It is recognized that these guidelines may not be appropriate for certain employers' circumstances and/or applications. In developing a written practice as required in Section 5, the employer shall review the detailed recommendations presented herein and shall modify them, as necessary, to meet particular needs.

2. Definitions

2.1 Terms included in this document are defined as follows:

(1) Certification: written testimony of qualification.

(2) Certifying Agency: the employer of the personnel being certified.

(3) Employer: the corporate, private, or public entity which employs personnel for wages, salary, fees, or other considerations.

(4) Experience: work activities accomplished in a specific NDT method under the direction of qualified supervision including performing the NDT method and related activities but *not* including time spent in organized training programs.

(5) Outside Agency: a company or individual who provides NDT Level III services and whose qualifications to provide these services have been reviewed by the employer engaging the company or individual.

(6) Qualification: demonstrated skill and knowledge and documented training and experience required for personnel to properly perform the duties of a specific job.

(7) Recommended Practice: a set of guidelines to assist the employer in developing uniform procedures for the qualification and certification of NDT personnel to satisfy the employer's specific requirements.

(8) Shall: a verb used to express the *minimum recommended guidelines* for most employers' qualification and certification programs; this is *not* intended to express what is mandatory by law or regulation.

(9) Should: a verb used to express the *desired guidelines* for most employers' qualification and certification programs.

(10) Training: the organized program developed to impart the knowledge and skills necessary for qualification.

3. Nondestructive Testing Methods

3.1 Qualification and certification of NDT personnel in accordance with this Recommended Practice is applicable to each of the following methods:

Radiographic Testing	RT
Magnetic Particle Testing	MT
Ultrasonic Testing	UT
Liquid Penetrant Testing	PT
Electromagnetic Testing	ET
Neutron Radiographic Testing	NRT
Leak Testing	LT
Acoustic Emission Testing	AET
Visual Testing	VT

4. Levels of Qualification

4.1 There are three basic levels of qualification. These levels may be further subdivided by the employer for specific situations where additional levels of skills and responsibilities are deemed necessary.

4.2 While in the process of being trained, qualified, and certified to NDT Level I, an individual should be considered a trainee. A trainee should work with a certified individual and shall not independently conduct any NDT, interpret or evaluate the results of any NDT, or report NDT results.

4.3 The three basic levels of qualification are as follow:

(1) **NDT Level I.** An NDT Level I individual shall be qualified to properly perform specific calibrations, specific NDT, and specific evaluations for acceptance or rejection determinations according to written instruc-

tions and to record results. The NDT Level I shall receive the necessary instruction or supervision from a certified NDT Level III individual or designee.

(2) **NDT Level II.** An NDT Level II individual shall be qualified to set up and calibrate equipment and to interpret and evaluate results with respect to applicable codes, standards, and specifications. The NDT Level II shall be thoroughly familiar with the scope and limitations of the methods for which qualified and should exercise assigned responsibility for on-the-job training and guidance of trainees and NDT Level I personnel. The NDT Level II shall be able to organize and report the results of NDT.

(3) **NDT Level III.** An NDT Level III individual shall be capable of establishing techniques and procedures; interpreting codes, standards, specifications, and procedures; and designating the particular NDT methods, techniques, and procedures to be used. The NDT Level III should be responsible for the NDT operations for which qualified and assigned and shall be capable of interpreting and evaluating results in terms of existing codes, standards, and specifications. The NDT Level III should have sufficient practical background in applicable materials, fabrication, and product technology to establish techniques and to assist in establishing acceptance criteria when none are otherwise available. The NDT Level III shall have general familiarity with other appropriate NDT methods, as demonstrated by the ASNT Level III Basic examination or other means. The NDT Level III, in the methods in which certified, should be capable of training and examining NDT Level I and II personnel for certification in those methods.

5. Written Practice

5.1 The employer shall establish a written practice for the control and administration of NDT personnel training, examination, and certification.

5.2 The employer's written practice shall describe the responsibility of each level of certification for determining the acceptability of materials or components in accordance with the applicable codes, standards, specifications, and procedures.

6. Education, Training, and Experience Requirements for Initial Qualification

6.1 Personnel considered for certification in NDT shall have sufficient education, training, and experience to ensure understanding of the principles and procedures of those NDT methods in which they are being considered for certification.

6.2 Documented training and/or experience gained in positions and activities equivalent to those of Levels I, II, and/or III prior to establishment of the employer's written practice may be considered in satisfying the criteria of Section 6.3.

6.3 To be considered for certification, a candidate shall satisfy one of the following criteria for the applicable NDT level:

(1) NDT Levels I and II

Table 6.3.1 (page 7) lists the recommended training and experience factors to be considered by the employer in establishing written practices for initial qualification of Level I and Level II individuals.

(2) NDT Level III

(a) Have graduated from a minimum four-year college or university curriculum with a degree in engineering or science, plus one year's experience in NDT testing in an assignment comparable to that of an NDT Level II in the applicable NDT method(s),

or:

(b) Have completed with passing grades at least two years of engineering or science study at a university, college, or technical school, plus two years' experience in assignments at least comparable to that of NDT Level II in the applicable NDT method(s),

or:

(c) Have four years' experience in an assignment at least comparable to that of an NDT Level II in the applicable NDT method(s).

When the individual is qualified by examination, the above requirements may be partially replaced by experience as a certified NDT Level II or by assignments at least comparable to NDT Level II in other methods listed in Section 3 of this Recommended Practice as defined in the employer's written practice.

7. Training Programs

7.1 Personnel being considered for initial certification shall complete sufficient organized training to become thoroughly familiar with the principles and practices of the specified NDT method related to the level of certification desired and applicable to the processes to be used and the products to be tested.

7.2 The training program shall include sufficient examinations to assure that the necessary information has been comprehended.

7.3 Recommended training course outlines for Levels I and II personnel and recommended references which may be used as technical source material follow in the Recommended Training Courses section of this Recommended Practice.

8. Examinations

8.1 An NDT Level III individual or a designated representative shall administer and grade the examinations specified in Section 8.3 through 8.8.

8.2 Vision Examinations

(1) Near-Vision Acuity. The examination shall assure natural or corrected near-distance acuity in at least one eye such that the applicant is capable of reading a minimum of Jaeger Number 2 or equivalent type and size letter at a distance of not less than 12 inches (30.5 cm) on a standard Jaeger test chart. The ability to perceive an Ortho-Rater minimum of 8 or similar test pattern is also acceptable. This shall be administered annually.

(2) Color Contrast Differentiation. The examination should demonstrate the capability of distinguishing and differentiating contrast among colors used in the method. This shall be conducted upon initial certification and at three-year intervals thereafter.

8.3 General (Written—for NDT Levels I and II)

(1) The general examinations shall address the basic principles of the applicable method.

(2) In preparing the examinations, the NDT Level III shall select or devise appropriate questions covering the applicable method to the degree required by the employer's written practice.

(3) See the Appendix for example questions.

(4) The number of questions which shall be given is as follows:

Test Method	Number of Level I Questions	Number of Level II Questions
Radiographic Testing	40	40
Magnetic Particle Testing	30	30
Ultrasonic Testing	40	40
Liquid Penetrant Testing	30	30
Electromagnetic Testing	40	40
Neutron Radiographic Testing	40	40
Leak Testing	20	20
Acoustic Emission Testing	40	40
Visual Testing	30	30

8.4 Specific (Written—for NDT Levels I and II)

(1) The specific examination shall address the equipment, operating procedures, and NDT techniques that the individual may encounter during specific assignments to the degree required by the employer's written practice.

(2) The specific examination should also cover the specifications or codes and acceptance criteria used in the employer's NDT procedures.

(3) The number of questions which shall be given is as follows:

Test Method	Number of Level I Questions	Number of Level II Questions
Radiographic Testing	20	20
Magnetic Particle Testing	20	15
Ultrasonic Testing	20	20
Liquid Penetrant Testing	20	15
Electromagnetic Testing	20	20
Neutron Radiographic Testing	15	15
Leak Testing		
1. Bubble Test	15	15
2. Absolute Pressure Leak Test (Pressure Change)	15	15
3. Halogen Diode Leak Test	15	15
4. Mass Spectrometer Leak Test	20	40
Acoustic Emission Testing	20	20
Visual Testing	20	20

8.5 Practical (for NDT Level I and II)

(1) The candidate should demonstrate familiarity with and ability to operate the necessary NDT equipment, record, and analyze the resultant information to the degree required.

(2) At least one selected specimen should be tested and the results of the NDT analyzed by the candidate.

(3) The description of the specimen, the NDT procedure, including check points, and the results of the examination shall be documented.

(4) NDT Level I Practical Examination. Proficiency shall be demonstrated in performing the applicable NDT on one or more samples approved by the NDT Level III. At least ten (10) different checkpoints requiring an understanding of test variables and the employer's procedural requirements shall be included in this practical examination.

(5) NDT Level II Practical Examination. Proficiency shall be demonstrated in selecting and performing the applicable NDT method and interpreting and evaluating the results on one or more samples approved by the NDT Level III. At least ten (10) different checkpoints requiring an understanding of NDT variables and the employer's procedural requirements shall be included in this practical examination.

8.6 Example examination questions for general examinations are presented in the separate question booklets which can be obtained from ASNT Headquarters. These questions are intended as examples only and should not be used verbatim for qualification examinations. The following is a list of the booklets:

Test Method	Question Booklets
Radiographic Test Method	A
Magnetic Particle Method	B
Ultrasonic Test Method	C
Liquid Penetrant Method	D
Electromagnetic (Eddy Current and Flux Leakage) Test Method	E
Neutron Radiographic Test Method	F
Acoustic Emission Test Method	G*
Leak Testing Method:	
Bubble Leak Testing	HB
Pressure Change Measurement	HP
Halogen Diode Detector Leak Testing	HH
Mass Spectrometer Test Method	HM
Visual Test Method	I*

*In course of preparation

8.7 The written examinations shall be administered without access to reference material (closed-book) except that necessary data, such as graphs, tables, specifications, procedures, codes, etc., may be provided. All questions used for Level I and Level II examinations shall be approved by the responsible Level III.

8.8 NDT Level III Examinations
(1) **Basic Examination** (required only once when more than one method examination is taken). The number of questions which shall be given is as follows:
  (a) Fifteen (15) questions relating to understanding the SNT-TC-1A document.
  (b) Twenty (20) questions relating to applicable materials, fabrication, and product technology.
  (c) Twenty (20) questions that are similar to published Level II questions for other appropriate NDT methods.
(2) **Method Examination** (for each method).
  (a) Thirty (30) questions relating to fundamentals and principles that are similar to published ASNT Level III questions for each method, and
  (b) Fifteen (15) questions relating to application and establishment of techniques and procedures that are similar to the published ASNT Level III questions for each method, and
  (c) Twenty (20) questions relating to capability for interpreting codes, standards, and specifications relating to the method.
(3) **Specific Examination** (for each method).
  (a) Twenty (20) questions relating to specifications, equipment, techniques, and procedures applicable to the employer's product(s) and methods employed and to the administration of the employer's written practice.
(4) A valid endorsement on an ASNT NDT Level III certificate fulfills the examination criteria described in 8.8(1) and 8.8(2) for each applicable NDT method.
(5) The employer may delete the specific examination described in 8.8(3)(a) if the candidate has a valid ASNT NDT Level III certificate in the method and if documented evidence of experience exists, including the preparation of NDT procedures to codes, standards, or specifications and the evaluation of test results.

8.9 Grading
(1) An NDT Level III shall be responsible for the administration and grading of examinations for NDT Level I and II personnel. The administration and grading of examinations may be delegated to a designated representative of the NDT Level III and so recorded. The actual administration and grading of Level III examinations may be performed by a designated representative of the employer.
(2) For Level I and II personnel, a composite grade shall be determined by simple averaging of the general, specific, and practical examination results. For Level III personnel, the composite grade shall be determined by simple averaging of the basic, method, and specific examination results.
(3) Examinations administered for qualification shall result in a composite grade of at least 80 percent, with no individual examination having a grade less than 70 percent.
(4) When an examination is administered and graded for the employer by an outside agency and the outside agency issues grades of pass or fail only, on a certified report, then the employer may accept the pass grade as 80 percent for that particular examination.

8.10 Reexamination
Those failing to attain the required grades shall wait at least thirty (30) days or receive suitable additional training as determined by the NDT Level III before reexamination.

9. Certification

9.1 Certification of all levels of NDT personnel is the responsibility of the employer.

9.2 The employer shall establish a written practice covering all phases of certification, including training, as specified in Section 5.

9.3 Certification of NDT personnel shall be based on demonstration of satisfactory qualification in accordance with Sections 6, 7, and 8, as modified by the employer's written practice.

9.4 At the option of the employer, an outside agency may be engaged to provide NDT Level III services. In such instances, the responsibility of certification is retained by the employer.

9.5 The employer who purchases outside services is responsible for assuring that the training and examination services are in accordance with the employer's written practice.

9.6 Personnel certifications and copies of the employer's written practice shall be maintained on file by the employer.

    (1) The qualification records of the certified individuals shall be maintained by the employer and shall include the following:

        (a) Name of certified individual.

        (b) Level of certification and NDT method.

        (c) Educational background and experience of certified individuals.

        (d) Statement indicating satisfactory completion of training in accordance with the employer's written practice.

        (e) Results of the vision examinations prescribed in 8.2 for the current certification period.

        (f) Current examination copy(ies) or evidence of successful completion of examinations.

        (g) Other suitable evidence of satisfactory qualifications when such qualifications are used in lieu of examinations.

        (h) Composite grade(s) or suitable evidence of grades.

        (i) Dates of certification and/or recertification and the dates of assignments to NDT.

        (j) Signature of employer's designated representative.

9.7 Recertification

    (1) All levels of NDT personnel shall be recertified periodically in accordance with one of the following criteria:

        (a) Evidence of continuing satisfactory performance.

        (b) Reexamination in those portions of the examinations in Section 8 deemed necessary by the employer's NDT Level III.

    (2) Recommended maximum recertification intervals are

        (a) Levels I and II - 3 years

        (b) Level III - 5 years

    (3) NDT personnel may be reexamined any time at the discretion of the employer and have their certifications extended or revoked.

    (4) The employer's written practice shall include rules covering the duration of interrupted service that requires reexamination and recertification.

10. Termination

10.1 The employer's certification shall be deemed revoked when employment is terminated.

10.2 A Level I, Level II, or Level III whose certification has been terminated may be certified to the former NDT level by a new employer based on examination, as described in Section 8, provided all of the following conditions are met to the new employer's satisfaction:

    (1) The employee has proof of prior certification.

    (2) The employee was working in the capacity to which certified within six (6) months of termination.

    (3) The employee is being recertified within six (6) months of termination.

Table 6.3.1

Recommended Initial Training and Experience Levels

Examination Method	RT		MT		UT		PT		ET		NRT		AET		VT		LT (I)				LT (II)			
Level	I	II	I	II	I	II	I	II	I	II	I	II	I	II	I	II								
Technique																	BT	PCT	HDLT	MSLT	BT	PCT	HDLT	MSLT
Initial Training (Hours) — A High school graduation or equivalent	39	40	12	8	40	40	4	8	40	40	28	40	40	40	8	16	2	24	12	40	4	16	8	24
B Completion with a passing grade of at least 2 years of engineering or science study in a university, college, or technical school	29	35	8	4	30	40	4	4	24	40	20	40	32	40	4	8	2	16	8	28	2	12	6	16
Experience Level (Months per Level)** — Educational levels as listed above	3	9	1	3	3	9	1	2	1	9	6	24	6	18	1	2	*	1½	1½	4	½	4	4	6

BT   = Bubble Test
PCT   = Pressure Change Test
HDLT = Halogen Diode Leak Test
MSLT = Mass Spectrometer Leak Test
*   = 2 Hours
**   = Work Time Experience per Level

Notes:

(1) For Level II certification, the experience shall consist of time at Level I or equivalent. If a person is being qualified directly to Level II with no time at Level I, the required experience shall consist of the sum of the times required for Level I and Level II and the required training shall consist of the sum of the hours required for Level I and Level II.

(2) Listed training hours may be adjusted as described in the employer's written practice depending on the candidate's actual education level, e.g., grammar school, college graduate in engineering, etc.

(3) Initial experience may be gained simultaneously in two or more methods if:

    (a) The candidate spends a mimimum of 25% of work time on each method for which certification is sought, and

    (b) The remainder of work time claimed as experience is spent in NDT-related activities defined in the employer's written practice.

(4) Training shall be outlined in the employer's written practice.

Recommended Training Courses

Radiographic Testing Method
(RT — Training Course Outline — TC-1)

	Recommended Hours of Instruction	
	A*	B*
Level I		
Radiographic Equipment Operating and Emergency Instructions Course	4	4
Basic Radiographic Physics Course	20	15
Radiographic Technique Course	15	10
Total	39	29
Level II		
Film Quality and Manufacturing Processes Course	20	15
Radiographic Evaluation and Interpretation Course	20	20
Total	40	35

*A High school graduate or equivalent.

*B Completion with passing grades of at least two years of engineering or science study at a university, college, or technical school.

This is a progressive training course; i.e., consideration as Level I is based on satisfactory completion of the Level I training course; consideration as Level II is based on satisfactory completion of both Level I and Level II training courses.

Topics in the training outline may be deleted or expanded to meet the employer's specific applications or for limited certification and may be accompanied by a corresponding change in training hours.

Recommended Training for Level I Radiographic Testing

Radiographic Equipment Operating and Emergency Instructions Course

Note: It is recommended that the trainee receive instruction in this course prior to performing work in radiography.

1. **Personnel Monitoring**
   a. Wearing of monitoring badges
   b. Reading of pocket dosimeters
   c. Recording of daily dosimeter readings
   d. "Off-scale" dosimeter—action required
   e. Permissible exposure limits

2. **Survey Instruments**
   a. Types of radiation instruments
   b. Reading and interpreting meter indications
   c. Calibration frequency
   d. Calibration expiration—action
   e. Battery check—importance

3. **Leak Testing of Sealed Radioactive Sources**
   a. Requirements for leak testing
   b. Purpose of leak testing
   c. Performance of leak testing

4. **Radiation Survey Reports**
   a. Requirements for completion
   b. Description of report format

5. **Radiographic Work Practices**
   a. Establishment of restricted areas
   b. Posting and surveillance of restricted areas
   c. Use of time, distance, and shielding to reduce personnel radiation exposure
   d. Applicable regulatory requirements for surveys, posting, and control of radiation and high-radiation areas

6. **Exposure Devices**
   a. Daily inspection and maintenance
   b.* Radiation exposure limits for gamma-ray exposure devices
   c. Labeling
   d. Use

   e. Use of collimators to reduce personnel exposure
   f.* Use of "source changers" for gamma-ray sources

7. **Emergency Procedures**
   a.* Vehicle accidents with radioactive sealed sources
   b.* Fire involving sealed sources
   c.* "Source out"—failure to return to safe shielded conditions
   d.* Emergency call list

8. **Storage and Shipment of Exposed Devices and Sources**
   a.* Vehicle storage
   b.* Storage vault—permanent
   c.* Shipping instructions—sources
   d.* Receiving instructions—radioactive material

9. **State and Federal Regulations**
   a. Nuclear Regulatory Commission (NRC) and agreement states—authority
   b. License reciprocity
   c.* Radioactive materials license requirements for industrial radiography
   d. Qualification requirements for radiography personnel
   e. Regulations for the control of radiation (state or NRC as applicable)
   f.* Department of Transportation regulations for radiographic-source shipment
   g. Regulatory requirements for X-ray machines (state and federal as applicable)

*Topics may be deleted if the radiography is limited to X-ray exposure devices.

Total recommended hours of instruction for this course:
   Classification A - 4 hours
   Classification B - 4 hours

Basic Radiographic Physics Course

1. **Introduction**
   a. History and discovery of radioactive materials
   b. Definition of industrial radiography
   c. Radiation protection—why?
   d. Basic math review: exponents, square root, etc.

2. **Fundamental Properties of Matter**
   a. Elements and atoms

   b. Molecules and compounds
   c. Atomic particles—properties of protons, electrons, and neutrons
   d. Atomic structure
   e. Atomic number and weight
   f. Isotope vs. radioisotope

3. **Radioactive Materials**
   a. Production
      (1) Neutron activation

    (2)   Nuclear fission
b. Stable vs. unstable (radioactive) atoms
c. Curie—the unit of activity
d. Half-life of radioactive materials
e. Plotting of radioactive decay
f. Specific activity—curies/gram

4. **Types of Radiation**
    a. Particulate radiation—properties: alpha, beta, neutron
    b. Electromagnetic radiation—X-ray, gamma ray
    c. X-ray production
    d. Gamma-ray production
    e. Gamma-ray energy
    f. Energy characteristics of common radioisotope sources
    g. Energy characteristics of X-ray machines

5. **Interaction of Radiation with Matter**
    a. Ionization
    b. Radiation interaction with matter
        (1)   Photoelectric effect
        (2)   Compton scattering
        (3)   Pair production
    c. Unit of radiation exposure—the roentgen
    d. Emissivity of commonly used radiographic sources
    e. Emissivity of X-ray exposure devices
    f. Attenuation of electromagnetic radiation—shielding
    g. Half-value layers; tenth-value layers
    h. Inverse-square law

6. **Biological Effects of Radiation**
    a. "Natural" background radiation
    b. Unit of radiation dose—rem
    c. Difference between radiation and contamination
    d. Allowable personnel-exposure limits and the banking concept
    e. Theory of allowable dose
    f. Radiation damage—repair concept
    g. Symptoms of radiation injury
    h. Acute radiation exposure and somatic injury
    i. Personnel monitoring for tracking exposure
    j. Organ radiosensitivity

7. **Radiation Detection**
    a. Pocket dosimeter
    b. Difference between dose and dose rate
    c. Survey instruments
        (1)   Geiger-Müller tube
        (2)   Ionization chambers
        (3)   Scintillation chambers, counters
    d. Film badge—radiation detector
    e. TLDs (thermoluminescent dosimeters)
    f. Calibration

8. **Exposure Devices and Radiation Sources**
    a. Radioisotope sources
        (1)   Sealed-source design and fabrication
        (2)   Gamma-ray sources
        (3)   Beta and bremsstrahlung sources
        (4)   Neutron sources
    b. Radioisotope exposure device characteristics
    c. Electronic radiation sources—500 keV and less, low-energy
        (1)   Generator—high-voltage rectifiers
        (2)   X-ray tube design and fabrication
        (3)   X-ray control circuits
        (4)   Accelerating potential
        (5)   Target material and configuration
        (6)   Heat dissipation
        (7)   Duty cycle
        (8)   Beam filtration
    d.* Electronic radiation sources—medium- and high-energy
        (1)* Resonance transformer
        (2)* Van de Graaff accelerator
        (3)* Linac
        (4)* Betatron
        (5)* Roentgen output
        (6)* Equipment design and fabrication
        (7)* Beam filtration
    e.* Fluoroscopic radiation sources
        (1)* Fluoroscopic equipment design
        (2)* Direct-viewing screens
        (3)* Image amplification
        (4)* Special X-ray tube considerations and duty cycle
        (5)* Screen unsharpness
        (6)* Screen conversion efficiency

9. **Special Radiographic Sources and Techniques**
    a.* Flash radiography
    b.* Stereo radiography
    c.* In-motion radiography
    d.* Autoradiography

*Topics may be deleted if these methods and techniques are not used by the employer.

Total recommended hours of instruction for this course:
    Classification A - 20 hours
    Classification B - 15 hours

Radiographic Technique Course

1. **Introduction**
   a. Process of radiography
   b. Types of electromagnetic radiation sources
   c. Electromagnetic spectrum
   d. Penetrating ability or "quality" of X-rays and gamma rays
   e. Spectrum of X-ray tube source
   f. Spectrum of gamma-radioisotope source
   g. X-ray tube—change of ma or kVp effect on "quality" and intensity

2. **Basic Principles of Radiography**
   a. Geometric exposure principles
      (1) "Shadow" formation and distortion
      (2) Shadow enlargement calculation
      (3) Shadow sharpness
      (4) Geometric unsharpness
      (5) Finding discontinuity depth
   b. Radiographic screens
      (1) Lead intensifying screens
      (2) Fluorescent intensifying screens
      (3) Intensifying factors
      (4) Importance of screen-to-film contact
      (5) Importance of screen cleaniness and care
      (6) Techniques for cleaning screens
   c. Radiographic cassettes
   d. Composition of industrial radiographic film
   e. The "heel effect" with X-ray tubes

3. **Radiographs**
   a. Formation of the latent image on film
   b. Inherent unsharpness
   c. Arithmetic of radiographic exposure
      (1) Milliamperage—distance-time relationship
      (2) Reciprocity law
      (3) Photographic density
      (4) X-ray exposure charts—material thickness, kV, and exposure
      (5) Gamma-ray exposure chart
      (6) Inverse-square-law considerations
      (7) Calculation of exposure time for gamma- and X-ray sources

   d. Characteristic Hurter and Driffield (H&D) curve
   e. Film speed and class descriptions
   f. Selection of film for particular purpose

4. **Radiographic Image Quality**
   a. Radiographic sensitivity
   b. Radiographic contrast
   c. Film contrast
   d. Subject contrast
   e. Definition
   f. Film graininess and screen mottle effects
   g. Penetrameters or image-quality indicators

5. **Film Handling, Loading, and Processing**
   a. Safe light and darkroom practices
   b. Loading bench and cleanliness
   c. Opening of film boxes and packets
   d. Loading of film and sealing cassettes
   e. Handling techniques for "green film"
   f. Elements of manual film processing

6. **Exposure Techniques—Radiography**
   a. Single-wall radiography
   b. Double-wall radiography
      (1) Viewing two walls simultaneously
      (2) Offset double-wall exposure single-wall viewing
      (3) Elliptical techniques
   c. Panoramic radiography
   d. Use of multiple-film loading
   e. Specimen configuration

7. **Fluoroscopic Techniques**
   a. Dark adaptation and eye sensitivity
   b. Special scattered radiation techniques
   c. Personnel protection
   d. Sensitivity
   e. Limitations
   f. Direct screen viewing
   g. Indirect and remote screen viewing

Total recommended hours of instruction for the course:
   Classification A - 15 hours
   Classification B - 10 hours

Recommended Training for Level II Radiographic Testing

Film Quality and Manufacturing Processes Course

1. **Review of Basic Radiographic Principles**
   a. Interaction of radiation with matter
   b. Math review
   c. Exposure calculations
   d. Geometric exposure principles
   e. Radiographic-image quality parameters

2. **Darkroom Facilities, Techniques, and Processing**
   a. Facilities and equipment
      (1) Automatic film processor vs. manual processing
      (2) Safe lights
      (3) Viewer lights
      (4) Loading bench
      (5) Miscellaneous equipment
   b. Film loading
      (1) General rules for handling unprocessed film
      (2) Types of film packaging
      (3) Cassette-loading techniques for sheet and roll
   c. Protection of radiographic film in storage
   d. Processing of film—manual
      (1) Developer and replenishment
      (2) Stop bath
      (3) Fixer and replenishment
      (4) Washing
      (5) Prevention of water spots
      (6) Drying
   e. Automatic film processing
   f. Film filing and storage
      (1) Retention-life measurements
      (2) Long-term storage
      (3) Filing and separation techniques
   g. Unsatisfactory radiographs—causes and cures
      (1) High film density
      (2) Insufficient film density
      (3) High contrast
      (4) Low contrast
      (5) Poor definition
      (6) Fog
      (7) Light leaks
      (8) Artifacts
   h. Film density
      (1) Step-wedge comparison film
      (2) Densitometers

3. **Indications, Discontinuities, and Defects**
   a. Indications
   b. Discontinuities
      (1) Inherent
      (2) Processing
      (3) Service
   c. Defects

4. **Manufacturing Processes and Associated Discontinuities**
   a. Casting processes and associated discontinuities
      (1) Ingots, blooms, and billets
      (2) Sand casting
      (3) Centrifugal casting
      (4) Investment casting
   b. Wrought processes and associated discontinuities
      (1) Forgings
      (2) Rolled products
      (3) Extruded products
   c. Welding processes and associated discontinuities
      (1) Submerged arc welding (SAW)
      (2) Shielded metal arc welding (SMAW)
      (3) Gas metal arc welding (GMAW)
      (4) Flux corded arc welding (FCAW)
      (5) Gas tungsten arc welding (GTAW)
      (6) Resistance welding
      (7) Special welding processes — electron beam, electroslag, electrogas, etc.

5. **Radiological Safety Principles Review**
   a. Controlling personnel exposure
   b. Time, distance, shielding concepts
   c. ALARA (as low as reasonably achievable) concept
   d. Radiation-detection equipment
   e. Exposure-device operating characteristics

Total recommended hours of instruction for this course:
   Classification A - 20 hours
   Classification B - 15 hours

Radiographic Evaluation and Interpretation Course

1. **Radiographic Viewing**
   a. Film-illuminator requirements
   b. Background lighting
   c. Multiple-composite viewing
   d. Penetrameter placement
   e. Personnel dark adaptation and visual acuity
   f. Film identification
   g. Location markers
   h. Film-density measurement
   i. Film artifacts

2. **Application Techniques**
   a. Multiple-film techniques
      (1) Thickness-variation parameters
      (2) Film speed
      (3) Film latitude
   b. Enlargement and projection
   c. Geometrical relationships
      (1) Geometrical unsharpness
      (2) Penetrameter sensitivity
      (3) Source-to-film distance
      (4) Focal-spot size
   d. Triangulation methods for discontinuity location
   e. Localized magnification
   f. Film-handling techniques

3. **Evaluation of Castings**
   a. Casting-method review
   b. Casting discontinuities

   c. Origin and typical orientation of discontinuities
   d. Radiographic appearance
   e. Casting codes/standards—applicable acceptance criteria
   f. Reference radiographs

4. **Evaluation of Weldments**
   a. Welding-method review
   b. Welding discontinuities
   c. Origin and typical orientation of discontinuities
   d. Radiographic appearance
   e. Welding codes/standards—applicable acceptance criteria
   f. Reference radiographs or pictograms

5. **Standards, Codes, and Procedures for Radiography**
   a. ASTM E94/E142
   b. Acceptable radiographic techniques and setups
   c. Applicable employer procedures
   d. Procedure for radiograph parameter verification
   e. Radiographic reports

Total recommended hours of instruction for this course:
   Classification A - 20 hours
   Classification B - 20 hours

Recommended Training References

Radiographic Testing Method

American Society for Metals. *Nondestructive Inspection and Quality Control: Metals Handbook,* Vol. 11, 8th ed. Metals Park, OH, 1976.*

American Society for Nondestructive Testing. *Question and Answer Book A: Radiographic Test Method, Levels I, II, and III.* Columbus, OH, 1980.*

American Society for Testing and Materials. ''Metallography; Nondestructive Testing,'' Vol. 03.03. Philadelphia, PA, Latest Edition.*

American Welding Society. *Welding Handbook.* Vol. 1. Miami, FL, Latest Edition.

__________. *Welding Inspection.* Miami, FL, Latest Edition.

British Institute of Non-Destructive Testing. *Basic Metallurgy for Non-Destructive Testing.* Essex, England: W. H. Houldershaw, Ltd., 1981.

Bryant, Lawrence A., and Paul McIntire, eds. *Nondestructive Testing Handbook: Radiography and Radiation Testing,* Vol. 3, 2nd ed. Columbus, OH: American Society for Nondestructive Testing, 1985.*

Halmshaw, R. *Industrial Radiology Techniques.* London: Wykeharn Publications; Ltd., New York: Springer-Verlag, Inc., 1971.

__________. *Industrial Radiology: Theory and Practice.* London and New Jersey: Applied Science Publishers, 1982.*

__________. *Non-destructive Testing: Metallurgy and Materials Science Series.* London: Edward Arnold, 1987.*

__________. *Physics of Industrial Radiology.* New York: American Elsevier Publishing Co., 1966.*

Hartford Steam Boiler's Complete Radiography Workbook, 1st ed. Hartford, CT: The Hartford Steam Boiler Inspection and Insurance Co., 1983.

Industrial Radiography/Holography. Ridgefield Park, NJ: Agfa-Gevaert, Inc., 1986.*

McGonnagle, Warren J. *Nondestructive Testing,* 2nd ed. New York: Gordon and Breach, 1975.

McGuire, Stephen A., and Carol A. Peabody. *Working Safely in Gamma Radiography,* NUREG/BR-0024. Washington, DC: U.S. Government Printing Office, 1982.*

McMaster, Robert C., ed. *Nondestructive Testing Handbook,* 1st ed. Columbus, OH: American Society for Nondestructive Testing, 1959.*

Mix, Paul E. *Introduction to Nondestructive Testing: A Training Guide.* New York: John Wiley & Sons, 1987.*

Moore, Harry D., ed. *Materials and Processes for NDT Technology.* Columbus, OH: American Society for Nondestructive Testing, 1981.*

Munro, John J., III, and Francis E. Roy, Jr. *Gamma Radiography Radiation Safety Handbook.* Burlington, MA: Amersham Corp., 1986.*

NDT Terminology. Wilmington, DE: E. I. du Pont de Nemours and Co., 1981.

Radiographic Testing, Classroom Training Handbook (CT-6-6). San Diego, CA: General Dynamics/ Convair Division, 1967.*

Radiographic Testing, Programmed Instruction Handbook (PI-4-6). San Diego, CA: General Dynamics/ Convair Division, 1983.*

Radiography in Modern Industry, 4th ed. Rochester, NY: Eastman Kodak Co., 1980.*

Richardson, Harry D. *NDT Radiography Training Manual.* Wilmington, DE: E. I. du Pont de Nemours and Co., 1968 reprint.*

Sensitometric Properties of X-Ray Films. Rochester, NY: Eastman Kodak Co., 1974.

Thielsch, Helmut. *Defects and Failures in Pressure Vessels and Piping.* New York: Reinhold Publishing Corp., 1966.*

__________. *The Sense and Nonsense of Weld Defects.* Morton Grove, IL: Monticello Books, 1967.*

*Available from the American Society for Nondestructive Testing, Columbus, OH.

Magnetic Particle Testing Method
(MT — Training Course Outline — TC-2)

	Recommended Hours of Instruction	
	A*	B*
Level I	12	8
Level II	8	4

*A High school graduate or equivalent.

*B Completion with passing grades of at least two years of engineering or science study at a university, college, or technical school.

This is a progressive training course; i.e., consideration as Level I is based on satisfactory completion of the Level I training course; consideration as Level II is based on satisfactory completion of both Level I and Level II training courses.

Topics in the training outline may be deleted or expanded to meet the employer's specific applications or for limited certification and may be accompanied by a corresponding change in training hours.

Recommended Training for
Level I Magnetic Particle Testing

1. **Principles of Magnets and Magnetic Fields**
   a. Theory of magnetic fields
      (1) Earth's magnetic field
      (2) Magnetic fields around magnetized materials
   b. Theory of magnetism
      (1) Magnetic poles
      (2) Law of magnetism
      (3) Materials influenced by magnetic fields
         (a) Ferromagnetic
         (b) Paramagnetic
      (4) Magnetic characteristics of nonferrous materials
   c. Terminology associated with magnetic particle testing

2. **Characteristics of Magnetic Fields**
   a. Bar magnet
   b. Ring magnet

3. **Effect of Discontinuities of Materials**
   a. Surface cracks
   b. Scratches
   c. Subsurface defects

4. **Magnetization by Means of Electric Current**
   a. Circular field
      (1) Field around a straight conductor
      (2) Right-hand rule
      (3) Field in parts through which current flows
         (a) Long, solid, cylindrical, regular parts
         (b) Irregularly shaped parts
         (c) Tubular parts
         (d) Parts containing machined holes, slots, etc.
      (4) Methods of inducing current flow in parts
         (a) Contact plates
         (b) Prods
      (5) Discontinuities commonly discovered by circular fields
   b. Longitudinal field
      (1) Field produced by current flow in a coil
      (2) Field direction in a current-carrying coil
      (3) Field strength in a current-carrying coil
      (4) Discontinuities commonly discovered by longitudinal fields
      (5) Advantages of longitudinal magnetization
      (6) Disadvantages of longitudinal magnetization

5. **Selecting the Proper Method of Magnetization**
   a. Alloy, shape, and condition of part
   b. Type of magnetizing current
   c. Direction of magnetic field
   d. Sequence of operations
   e. Value of flux density

6. **Inspection Materials**
   a. Wet particles
   b. Dry particles

7. **Principles of Demagnetization**
   a. Residual magnetism
   b. Reasons for requiring demagnetization
   c. Longitudinal and circular residual fields
   d. Basic principles of demagnetization
   e. Retentivity and coercive force
   f. Methods of demagnetization

8. **Magnetic Particle Testing Equipment**
   a. Equipment-selection considerations
      (1) Type of magnetizing current
      (2) Location and nature of test
      (3) Test materials used
      (4) Purpose of test
      (5) Area inspected
   b. Manual inspection equipment
   c. Medium- and heavy-duty equipment
   d. Stationary equipment
   e. Mechanized inspection equipment
      (1) Semiautomatic inspection equipment
      (2) Single-purpose semiautomatic equipment
      (3) Multipurpose semiautomatic equipment
      (4) Fully automatic equipment

9. **Types of Discontinuities Detected by Magnetic Particle Testing**
   a. Inclusions
   b. Blowholes
   c. Porosity
   d. Flakes
   e. Cracks
   f. Pipes
   g. Laminations
   h. Laps
   i. Forging bursts
   j. Voids

10. **Magnetic Particle Test Indications and Interpretations**
   a. Indications of nonmetallic inclusions
   b. Indications of surface seams
   c. Indications of cracks
   d. Indications of laminations
   e. Indications of laps
   f. Indications of bursts and flakes
   g. Indications of porosity
   h. Nonrelevant indications

Total recommended hours of instruction for this course:

Classification A - 12 hours
Classification B - 8 hours

Recommended Training for Level II Magnetic Particle Testing

1. **Principles**
   a. Theory
      (1) Flux patterns
      (2) Frequency and voltage factors
      (3) Current calculations
      (4) Surface flux strength
      (5) Subsurface effects
   b. Magnets and magnetism
      (1) Distance factors vs. strength of flux
      (2) Internal and external flux patterns
      (3) Phenomenon action at the discontinuity
      (4) Heat effects on magnetism
      (5) Material hardness vs. magnetic retention

2. **Flux Fields**
   a. Direct current
      (1) Depth of penetration factors
      (2) Source of current
   b. Direct pulsating current
      (1) Similarity to direct current
      (2) Advantages
      (3) Typical fields
   c. Alternating current
      (1) Cyclic effects
      (2) Surface strength characteristics
      (3) Safety precautions
      (4) Voltage and current factors
      (5) Source of current

3. **Effects of Discontinuities on Materials**
   a. Design factors
      (1) Mechanical properties
      (2) Part use
   b. Relationship to load-carrying ability

4. **Magnetization by Means of Electric Current**
   a. Circular techniques
      (1) Current calculations
      (2) Depth-factor considerations
      (3) Precautions—safety and overheating
      (4) Contact prods and yokes
         (a) Requirements for prods and yokes
         (b) Current-carrying capabilities
      (5) Discontinuities commonly detected
   b. Longitudinal technique
      (1) Principles of induced flux fields
      (2) Geometry of part to be inspected
      (3) Shapes and sizes of coils
      (4) Use of coils and cables
         (a) Strength of field
         (b) Current directional flow vs. flux field
         (c) Shapes, sizes, and current capacities
      (5) Current calculations
         (a) Formulas
         (b) Types of current required
         (c) Current demand

      (6) Discontinuities commonly detected

5. **Selecting the Proper Method of Magnetization**
   a. Alloy, shape, and condition of part
   b. Type of magnetizing current
   c. Direction of magnetic field
   d. Sequence of operations
   e. Value of flux density

6. **Demagnetization Procedures**
   a. Need for demagnetization of parts
   b. Current, frequency, and field orientation
   c. Heat factors and precautions
   d. Need for collapsing flux fields

7. **Equipment**
   a. Portable type
      (1) Reason for portable equipment
      (2) Capabilities of portable equipment
      (3) Similarity to stationary equipment
   b. Stationary type
      (1) Capability of handling large and heavy parts
      (2) Flexibility in use
      (3) Need for stationary equipment
      (4) Use of accessories and attachments
   c. Automatic type
      (1) Requirements for automation
      (2) Sequential operations
      (3) Control and operation factors
      (4) Alarm and rejection mechanisms
   d. Liquids and powders
      (1) Liquid requirements as a particle vehicle
      (2) Safety precautions
      (3) Temperature needs
      (4) Powder and paste contents
      (5) Mixing procedures
      (6) Need for accurate proportions
   e. Black-light type
      (1) Black light and fluorescence
      (2) Visible- and black-light comparisons
      (3) Requirements in the testing cycle
      (4) Techniques in use
   f. Light-sensitive instruments
      (1) Need for instrumentation
      (2) Light characteristics

8. **Types of Discontinuities**
   a. In castings
   b. In ingots
   c. In wrought sections and parts
   d. In welds

9. **Evaluation Techniques**
   a. Use of standards
      (1) Need for standards and references
      (2) Comparison of known with unknown
      (3) Specifications and certifications
      (4) Comparison techniques
   b. Defect appraisal

(1) History of part
(2) Manufacturing process
(3) Possible causes of defect
(4) Use of part
(5) Acceptance and rejection criteria
(6) Use of tolerances

10. Quality Control of Equipment and Processes
   a. Malfunctioning of equipment
   b. Proper magnetic particles and bath liquid
   c. Bath concentration
      (1) Settling test
      (2) Other bath-strength tests
   d. Tests for black-light intensity

Total recommended hours of instruction for this course:

    Classification A - 8 hours
    Classification B - 4 hours

Recommended Training References

Magnetic Particle Testing Method

American Society for Metals. *Nondestructive Inspection and Quality Control: Metals Handbook*, Vol. 11, 8th ed. Metals Park, OH, 1976.*

American Society for Nondestructive Testing. *Question and Answer Book B: Magnetic Particle Test Method: Levels I, II and III*. Columbus, OH, 1980.*

American Society for Testing and Materials. "Metallography; Nondestructive Testing," Vol. 03.03. Philadelphia, PA, Latest Edition.*

American Welding Society. *Welding Handbook*. Vol. 1. Miami, FL, Latest Edition.

Betz, Carl E. *Principles of Magnetic Particle Testing*. Chicago: Magnaflux Corp., 1985.*

Magnetic Particle Testing, Classroom Training Handbook (CT-6-3). San Diego, CA: General Dynamics/Convair Division, 1977.*

Magnetic Particle Testing, Programmed Instruction Handbook (PI-4-3). San Diego, CA: General Dynamics/Convair Division, 1977.*

McGonnagle, Warren J. *Nondestructive Testing*, 2nd ed. New York: Gordon and Breach, 1975.

McMaster, Robert C., ed. *Nondestructive Testing Handbook*, 1st ed. Columbus, OH: American Society for Nondestructive Testing, 1959.*

Mix, Paul E. *Introduction to Nondestructive Testing: A Training Guide*. New York: John Wiley & Sons, 1987.*

*Available from the American Society for Nondestructive Testing, Columbus, OH.

Ultrasonic Testing Method
(UT — Training Course Outline — TC-3)

	Recommended Hours of Instruction	
	A*	B*
Level I		
Basic Ultrasonic Course	20	15
Ultrasonic Technique Course	20	15
Total	40	30
Level II		
Ultrasonic Evaluation Course	40	40

*A High school graduate or equivalent.

*B Completion with passing grades of at least two years of engineering or science study at a university, college, or technical school.

This is a progressive training course; i.e., consideration as Level I is based on satisfactory completion of the Level I training course; consideration as Level II is based on satisfactory completion of both Level I and Level II training courses.

Topics in the training outline may be deleted or expanded to meet the employer's specific applications or for limited certification and may be accompanied by a corresponding change in training hours.

Recommended Training for Level I Ultrasonic Testing

Basic Ultrasonic Course

Note: It is recommended that the trainee receive instruction in this course prior to performing work in ultrasonics.

1. **Introduction**
   a. Definition of ultrasonics
   b. History of ultrasonic testing
   c. Applications of ultrasonic energy
   d. Basic math review
   e. Responsibilities of levels of certification

2. **Basic Principles of Acoustics**
   a. Nature of sound waves
   b. Modes of sound-wave generation
   c. Velocity, frequency, and wavelength of sound waves
   d. Attenuation of sound waves
   e. Acoustic impedance
   f. Reflection
   g. Refraction and mode-conversion
   h. Snell's law and critical angles
   i. Fresnel and Fraunhofer effects

3. **Equipment**
   a. Basic pulse-echo instrumentation (A-, B-, and C-scan)
      (1) Electronics—time base, pulser, receiver, and cathode-ray tube (CRT)
      (2) Control functions
      (3) Calibration
          (a) Basic instrument calibration
          (b) Calibration blocks (types and use)
   b. Digital thickness instrumentation
   c. Transducer operation and theory
      (1) Piezoelectric effect
      (2) Types of crystals
      (3) Frequency (crystal-thickness relationships)
      (4) Nearfield and farfield
      (5) Beam spread
      (6) Construction, materials, and shapes
      (7) Types (straight, angle, dual, etc.)
      (8) Beam-intensity characteristics
      (9) Sensitivity, resolution, and damping
      (10) Mechanical vibration into part
   d. Couplants
      (1) Purpose and principles
      (2) Materials and their efficiency

4. **Basic Testing Methods**
   a. Contact
   b. Immersion

Total recommended hours of instruction for this course:
Classification A - 20 hours
Classification B - 15 hours

Ultrasonic Technique Course

1. **Testing Methods**
   a. Contact
      (1) Straight-beam
      (2) Angle-beam
      (3) Surface-wave
      (4) Pulse-echo transmission
      (5) Multiple transducer
      (6) Curved surfaces
   b. Immersion
      (1) Transducer in water
      (2) Water column, wheels, etc.
      (3) Submerged test part
      (4) Sound-beam path—transducer to part
      (5) Focused transducers
      (6) Curved surfaces
   c. Comparison of contact and immersion methods

2. **Calibration (Electronic and Functional)**
   a. Equipment
      (1) Cathode-ray tube (amplitude, sweep, etc.)
      (2) Recorders
      (3) Alarms
      (4) Automatic and semiautomatic systems
      (5) Electronic distance/amplitude correction
      (6) Transducers
   b. Calibration of equipment electronics
      (1) Variable effects
      (2) Transmission accuracy
      (3) Calibration requirements
      (4) Calibration reflectors
   c. Inspection calibration
      (1) Comparison with reference blocks
      (2) Pulse-echo variables
      (3) Reference for planned tests (straight-beam, angle-beam, etc.)
      (4) Transmission factors
      (5) Transducer
      (6) Couplants
      (7) Materials

3. **Straight-Beam Examination to Specific Procedures**
   a. Selection of parameters
   b. Test standards
   c. Evaluation of results
   d. Test reports

4. **Angle-Beam Examination to Specific Procedures**
   a. Selection of parameters
   b. Test standards
   c. Evaluation of results
   d. Test reports

Total recommended hours of instruction for this course:

    Classification A - 20 hours

    Classification B - 15 hours

Recommended Training for Level II Ultrasonic Testing

Ultrasonic Evaluation Course

1. **Review of Ultrasonic Technique Course**
   a. Principles of ultrasonics
   b. Equipment
   c. Testing techniques
   d. Calibration
      (1) Straight-beam
      (2) Angle-beam
      (3) Resonance
      (4) Special applications
2. **Evaluation of Base-Material Product Forms**
   a. Ingots
      (1) Process review
      (2) Types, origin, and typical orientation of discontinuities
      (3) Response of discontinuities to ultrasound
      (4) Applicable codes/standards
   b. Plate and sheet
      (1) Rolling process
      (2) Types, origin, and typical orientation of discontinuities
      (3) Response of discontinuities to ultrasound
      (4) Applicable codes/standards
   c. Bar and Rod
      (1) Forming process
      (2) Types, origin, and typical orientation of discontinuities
      (3) Response of discontinuities to ultrasound
      (4) Applicable codes/standards
   d. Pipe and tubular products
      (1) Manufacturing process
      (2) Types, origin, and typical orientation of discontinuities
      (3) Response of discontinuities to ultrasound
      (4) Applicable codes/standards
   e. Forgings
      (1) Process review
      (2) Types, origin, and typical orientation of discontinuities
      (3) Response of discontinuities to ultrasound
      (4) Applicable codes/standards
   f. Castings
      (1) Process review
      (2) Types, origin, and typical orientation of discontinuities
      (3) Response of ultrasound to discontinuities
      (4) Applicable codes/standards
   g. Other product forms as applicable—rubber, glass, etc.
3. **Evaluation of Weldments**
   a. Welding processes
   b. Weld geometries
   c. Welding discontinuities
   d. Origin and typical orientation of discontinuities
   e. Response of discontinuities to ultrasound
   f. Applicable codes/standards
4. **Evaluation of Bonded Structures**
   a. Manufacturing processes
   b. Types of discontinuities
   c. Origin and typical orientation of discontinuities
   d. Response of discontinuities to ultrasound
   e. Applicable codes/standards
5. **Discontinuity Detection**
   a. Sensitivity to reflections
      (1) Size, type, and location of discontinuities
      (2) Techniques used in detection
      (3) Wave characteristics
      (4) Material and velocity
      (5) Discontinuity
   b. Resolution
      (1) Standard reference comparisons
      (2) History of part
      (3) Probability of type of discontinuity
      (4) Degrees of operator discrimination
      (5) Effects of ultrasonic frequency
      (6) Damping effects
   c. Determination of discontinuity size
      (1) Cathode-ray tube (CRT) display and meter indications
      (2) Transducer movement vs. display
      (3) Two-dimensional testing techniques
      (4) Signal patterns
   d. Location of discontinuity
      (1) CRT display
      (2) Amplitude and linear time
      (3) Search technique
6. **Evaluation**
   a. Comparison procedures
      (1) Standards and references
      (2) Amplitude, area, and distance relationship
      (3) Application of results of other NDT methods
   b. Object appraisal
      (1) History of part
      (2) Intended use of part
      (3) Existing and applicable code interpretation
      (4) Type of discontinuity and location

Total recommended hours of instruction for this course:
Classification A - 40 hours
Classification B - 40 hours

Recommended Training References

Ultrasonic Testing Method

American Society for Metals. *Nondestructive Inspection and Quality Control: Metals Handbook*, Vol. 11, 8th ed. Metals Park, OH, 1976.*

American Society for Nondestructive Testing. *Question and Answer Book C: Ultrasonic Test Method, Levels I, II, and III*. Columbus, OH, 1980*

American Society for Testing and Materials. "Metallography; Nondestructive Testing," Vol. 03.03. Philadelphia, PA, Latest Edition.*

American Welding Society. *Welding Handbook*. Vol. 1. Miami, FL, Latest Edition.

Ensminger, D. *Ultrasonics: Fundamentals, Technology, Applications*, 2nd ed. New York and Basel: Marcel Dekker, Inc., 1988.*

__________. *Ultrasonics: The Low and High Intensity Applications*. New York: Marcel Dekker, Inc., 1973.

Halmshaw, R. *Non-destructive Testing: Metallurgy and Materials Science Series*. London: Edward Arnold, 1987.*

Handbook for Standardization of Nondestructive Testing Methods, MIL-HDBK-333 (USAF), Vol. 2. Washington, DC: U.S. Government Printing Office, 1974.

Krautkramer, Josef, and Herbert Krautkramer. *Ultrasonic Testing of Materials*, 3rd ed. New York: Springer-Verlag, 1983.*

McGonnagle, Warren J. *Nondestructive Testing*, 2nd ed. New York: Gordon and Breach, 1975.

McMaster, Robert C., ed. *Nondestructive Testing Handbook*, 1st ed. Columbus, OH: American Society for Nondestructive Testing, 1959.*

Mix, Paul E. *Introduction to Nondestructive Testing: A Training Guide*. New York: John Wiley & Sons, 1987.*

Procedures and Recommendations for the Ultrasonic Testing of Butt Welds, 2nd ed. London: The Welding Institute, 1972.*

Rose, J. L., and B. B. Goldberg. *Basic Physics in Diagnostic Ultrasound*. New York: John Wiley & Sons, 1979.*

Silvus, H. S., Jr. *Advanced Ultrasonic Testing Systems: A State of the Art Survey*. San Antonio, TX: Nondestructive Testing Information Analysis Center, 1977.*

Ultrasonic Method Training Program: Instructor's Package. Columbus, OH: American Society for Nondestructive Testing, 1981.*

Ultrasonic Method Training Program: Student's Package. Columbus, OH: American Society for Nondestructive Testing, 1981.*

Ultrasonic Testing, Classroom Training Handbook (CT-6-4). San Diego, CA: General Dynamics/Convair Division, 1967.*

Ultrasonic Testing, Programmed Instruction Handbook (PI-4-4), Vols. 1, 2, and 3. San Diego, CA: General Dynamics/Convair Division, 1967.*

*Available from the American Society for Nondestructive Testing, Columbus, OH.

Liquid Penetrant Testing Method
(LP — Training Course Outline — TC-4)

<table>
<tr><td></td><td colspan="2" align="center">Recommended Hours
of Instruction</td></tr>
<tr><td></td><td align="center">A*</td><td align="center">B*</td></tr>
<tr><td>Level I</td><td align="center">4</td><td align="center">4</td></tr>
<tr><td>Level II</td><td align="center">8</td><td align="center">4</td></tr>
</table>

*A High school graduate or equivalent.

*B Completion with passing grades of at least two years of engineering or science study at a university, college, or technical school.

This is a progressive training course; i.e., consideration as Level I is based on satisfactory completion of the Level I training course; consideration as Level II is based on satisfactory completion of both Level I and Level II training courses.

Topics in the training outline may be deleted or expanded to meet the employer's specific applications or for limited certification and may be accompanied by a corresponding change in training hours.

Recommended Training for Level I Liquid Penetrant Testing

1. **Introduction**
   a. Brief history of nondestructive testing and liquid penetrant testing
   b. Purpose of liquid penetrant testing
   c. Basic principles of liquid penetrant testing
   d. Types of liquid penetrants commercially available

2. **Liquid Penetrant Processing**
   a. Preparation of parts
   b. Adequate lighting
   c. Application of penetrant to parts
   d. Removal of surface penetrant
   e. Developer application and drying
   f. Inspection and evaluation
   g. Postcleaning

3. **Various Penetrant Testing Methods**
   a. Characteristics of each method
   b. General applications of each method

4. **Liquid Penetrant Testing Equipment**
   a. Liquid penetrant testing units
   b. Lighting for liquid penetrant inspection
   c. Materials for liquid penetrant testing
   d. Precautions in penetrant inspection

Total recommended hours of instruction for this course:
   Classification A - 4 hours
   Classification B - 4 hours

Recommended Training for Level II Liquid Penetrant Testing

1. **Review**
   a. Basic principles
   b. Process of various methods
   c. Equipment

2. **Selection of the Appropriate Penetrant Testing Method**
   a. Advantages of various methods
   b. Disadvantages of various methods

3. **Inspection and Evaluation of Indications**
   a. General
      (1) Discontinuities inherent in various materials
      (2) Reason for indications
      (3) Appearance of indications
      (4) Time for indications to appear
      (5) Persistence of indications
   b. Factors affecting indications
      (1) Penetrant used
      (2) Prior processing
      (3) Technique used
   c. Indications from cracks
      (1) Cracks occurring during solidification
      (2) Cracks occurring during processing
      (3) Cracks occurring during service
   d. Indications from porosity
   e. Indications from specific material forms
      (1) Forgings
      (2) Castings
      (3) Plate
      (4) Welds
      (5) Extrusions
   f. Evaluation of indications
      (1) True indications
      (2) False indications
      (3) Relevant indications
      (4) Nonrelevant indications

4. **Inspection Procedures and Standards**
   a. Inspection procedures
   b. Standards/codes

Total recommended hours of instruction for this course:

   Classification A - 8 hours
   Classification B - 4 hours

Recommended Training References

Liquid Penetrant Testing Method

American Society for Metals. *Nondestructive Inspection and Quality Control: Metals Handbook*, Vol. 11, 8th ed. Metals Park, OH, 1976.*

American Society for Nondestructive Testing. *Question and Answer Book D: Liquid Penetrant Test Method, Levels I, II, and III*. Columbus, OH, 1980.*

American Society for Testing and Materials. "Metallography; Nondestructive Testing," Vol. 03.03. Philadelphia, PA, Latest Edition.*

American Welding Society. *Welding Handbook*, Vol. 1. Miami, FL, Latest Edition.

Liquid Penetrant Testing, Classroom Training Handbook (CT-6-2). San Diego, CA: General Dynamics/ Convair Division, 1977.*

Liquid Penetrant Testing, Programmed Instruction Handbook (PI-4-2). San Diego, CA: General Dynamics/ Convair Division, 1977.*

Materials Evaluation, Vol. 44, No. 12 (November 1986). Columbus, OH: American Society for Nondestructive Testing.*

Materials Evaluation, Vol. 45, No. 7 (July 1987). Columbus, OH: American Society for Nondestructive Testing.*

McGonnagle, Warren J. *Nondestructive Testing*, 2nd ed. New York: Gordon and Breach, 1975.

McMaster, Robert C., ed. *Nondestructive Testing Handbook*, 1st ed. Columbus, OH: American Society for Nondestructive Testing, 1959.*

__________. *Nondestructive Testing Handbook: Liquid Penetrant Tests*, Vol. 2, 2nd ed. Columbus, OH: American Society for Nondestructive Testing, 1982.*

Mix, Paul E. *Introduction to Nondestructive Testing: A Training Guide*. New York: John Wiley & Sons, 1987.*

*Available from the American Society for Nondestructive Testing, Columbus, OH.

...gnetic Testing Method
...ing Course Outline — TC-5)

	Recommended Hours of Instruction	
	A*	B*
	24	12
	16	12
Total	40	24
Total	40	40

*A High school graduate or equivalent.

*B Completion with passing grades of at least two years of engineering or science study at a university, college, or technical school.

This is a progressive training course; i.e., consideration as Level I is based on satisfactory completion of the Level I training course; consideration as Level II is based on satisfactory completion of both Level I and Level II training courses.

Topics in the training outline may be deleted or expanded to meet the employer's specific applications or for limited certification and may be accompanied by a corresponding change in training hours.

Recommended Training for
Level I Electromagnetic Testing

Basic Electromagnetic Physics Course

1. **Introduction to Electromagnetic Testing (Eddy Current/Flux Leakage)**
   a. Brief history of testing
   b. Basic principles of testing

2. **Electromagnetic Theory**
   a. Eddy current theory
      (1) Generation of eddy currents by means of an AC field
      (2) Effect of fields created by eddy currents (impedance changes)
      (3) Effect of change of impedance on instrumentation
      (4) Properties of eddy current
         (a) Travel in circular direction
         (b) Strongest on surface of test material
         (c) Zero value at center of solid conductor placed in an alternating magnetic field
         (d) Strength, time relationship, and orientation as functions of test-system parameters and test-part characteristics
         (e) Have properties of compressible fluids
         (f) Small magnitude of current flow
         (g) Relationship of frequency and plane with current in coil
         (h) Effective permeability variations when induced in magnetic materials
         (i) Effect of discontinuity orientation
         (j) Power losses
   b. Flux leakage theory
      (1) Terminology and units
      (2) Principles of magnetization
         (a) B-H curve
         (b) Magnetic properties
         (c) Magnetic field
         (d) Hysteresis loop
         (e) Magnetic permeability
         (f) Factors affecting permeability
      (3) Magnetization—electromagnetism theory
         (a) Oersted's law
         (b) Faraday's law
         (c) Electromagnetics
      (4) Flux leakage theory and principle
         (a) Residual
         (b) Active
         (c) Tangential leakage
         (d) Normal leakage fields

Total recommended hours of instruction for this course:
   Classification A - 24 hours
   Classification B - 12 hours

Electromagnetic Technique Course

1. **Readout Mechanism**
   a. Calibrated or uncalibrated meter
   b. Null meter with dial indicator
   c. Oscilloscope
   d. Alarm, lights, etc.
   e. Numerical counters
   f. Marking system
   g. Sorting gates and tables
   h. Cutoff saw or shears
   i. Automation and feedback
   j. Strip-chart recorder

2. **Types of Eddy Current Sensing Elements**
   a. Probes
      (1) Types of arrangements
         (a) Absolute
         (b) Differential
      (2) Lift-off
      (3) Theory of operation
      (4) Applications
      (5) Advantages
      (6) Limitations
   b. Through, encircling, or annular coils
      (1) Types of arrangements
         (a) Absolute
         (b) Differential
      (2) Fill factor
      (3) Theory of operation
      (4) Applications
      (5) Advantages
      (6) Limitations
   c. Factors affecting choice of sensing elements
      (1) Type of part to be inspected
      (2) Type of discontinuity to be detected
      (3) Speed of testing required
      (4) Amount of testing (percentage) required
      (5) Probable location of discontinuity

3. **Types of Flux Leakage Sensing Elements**
   a. Principles of magnetic-measurement techniques
   b. Inductive-coil sensors
      (1) Theory of electromotive force (emf) induced in coil

   (2) Various constructions and designs of coils
   (3) Coil parameters affecting the flux leakage response
   (4) Sensing-coil systems and connections (single- and multielement probes)
c. Semiconductor sensing elements
   (1) Hall-effect probes
   (2) Magnetoresistors
   (3) Magnetodiodes
   (4) Magnetotransistors
   (5) Magnetic and electric characteristics of semiconductor sensing elements
d. Other methods of magnetic leakage field detection
   (1) Magnetic-tape system
   (2) Magnetic powder
   (3) Magnetic-resonance sensor

Total recommended hours of instruction for this course:

Classification A - 16 hours
Classification B - 12 hours

Recommended Training for Level II Electromagnetic Testing

Electromagnetic Evaluation Course

1. **Review of Electromagnetic Theory**
   a. Eddy current theory
   b. Flux leakage theory
   c. Types of eddy current sensing probes
   d. Types of flux leakage sensing probes

2. **Factors That Affect Coil Impedance**
   a. Test part
      (1) Conductivity
      (2) Permeability
      (3) Mass
      (4) Homogeneity
   b. Test system
      (1) Frequency
      (2) Coupling
      (3) Field strength
      (4) Test coil and shape

3. **Factors That Affect Flux Leakage Fields**
   a. Degree of magnetization
   b. Defect geometry
   c. Defect location
   d. Defect orientation
   e. Velocity factor
   f. Distance between adjacent defects

4. **Signal-to-Noise Ratio**
   a. Definition
   b. Relationship to eddy current testing
   c. Relationship to flux leakage testing
   d. Methods of improving signal-to-noise ratio

5. **Selection of Test Frequency**
   a. Relationship of frequency to type of test
   b. Considerations affecting choice of test
      (1) Signal-to-noise ratio
      (2) Phase discrimination
      (3) Response speed
      (4) Skin effect

6. **Selection of Method of Magnetization for Flux Leakage Testing**
   a. Magnetization characteristics for various magnetic materials
   b. Magnetization by means of electric fields
      (1) Circular field
      (2) Longitudinal field
      (3) Value of flux density
   c. Magnetization by means of permanent magnets
      (1) Permanent magnet relationship and theory
      (2) Permanent magnet materials

   d. Selection of proper magnetization method

7. **Coupling**
   a. "Fill factor" in through-coil inspection
   b. "Lift-off" and compensation in probe coil inspection
   c. Flux leakage "fill factor" in flux leakage testing
   d. "Lift-off" in flux leakage testing

8. **Field Strength and Its Selection**
   a. Permeability changes
   b. Saturation
   c. Effect of AC field strength on eddy current testing
   d. Effect of field strength in flux leakage testing

9. **Field Orientation for Flux Leakage Testing**
   a. Circular field
   b. Longitudinal field

10. **Instrument Design Considerations**
    a. Amplification
    b. Phase detection
    c. Differentiation of filtering

11. **Applications**
    a. Flaw detection
       (1) Eddy current methods
       (2) Flux leakage methods
    b. Sorting for properties related to conductivity—eddy current
    c. Sorting for properties related to permeability
       (1) Eddy current methods
       (2) Flux leakage methods
    d. Thickness evaluation—eddy current
    e. Measurement of magnetic-characteristic values
       (1) Eddy current methods
       (2) Flux leakage methods

12. **User Standards and Operating Procedures**
    a. Explanation of standards and specifications used in electromagnetic testing
    b. Explanation of operating procedures used in electromagnetic testing

Total recommended hours of instruction for this course:
   Classification A - 40 hours
   Classification B - 40 hours

Recommended Training References

Electromagnetic Testing Method

American Radio Relay League. *Radio Amateur's Handbook*. Neurington, CT.

American Society for Metals. *Nondestructive Inspection and Quality Control: Metals Handbook*, Vol. 11, 8th ed. Metals Park, OH, 1976.*

American Society for Nondestructive Testing. *Question and Answer Book E: Eddy Current Test Method and Flux Leakage Test Method, Levels I, II, and III.* Columbus, OH, 1980.*

American Society for Testing and Materials. "Metallography; Nondestructive Testing," Vol. 03.03. Philadelphia, PA, Latest Edition.*

Cecco, V. S., G. Van Drunen, and F. L. Sharp. *Eddy Current Testing*. Columbia, MD: GP Courseware, 1987.

Eddy Current Testing, Classroom Training Handbook (CT-6-5). San Diego, CA: General Dynamics/Convair Division, 1979.*

Eddy Current Testing, Programmed Instruction Handbook (PI-4-5). San Diego, CA: General Dynamics/Convair Division, 1980.*

Libby, Hugo L. *Introduction to Electromagnetic Nondestructive Test Methods*. Huntington, NY: Robert E. Krieger Publishing Co., 1971.*

McGonnagle, Warren J. *Nondestructive Testing*, 2nd ed. New York: Gordon and Breach, 1975.

McMaster, Robert C., ed. *Nondestructive Testing Handbook*, 1st ed. Columbus, OH: American Society for Nondestructive Testing, 1959.*

Mester, M. L., and Paul McIntire, eds. *Nondestructive Testing Handbook: Electromagnetic Testing*, Vol. 4, 2nd ed. Columbus, OH: American Society for Nondestructive Testing, 1986.*

Mix, Paul E. *Introduction to Nondestructive Testing: A Training Guide*. New York: John Wiley & Sons, 1987.*

*Available from the American Society for Nondestructive Testing, Columbus, OH.

Neutron Radiographic Testing Method
(NRT — Training Course Outline — TC-6)

	Recommended Hours of Instruction	
	A*	B*
Level I		
Neutron Radiographic Equipment Operating and Emergency Instructions Course	8	8
Basic Neutron Radiographic Physics Course	7	4
Basic Neutron Radiographic Technique Course	13	8
Total	28	20
Level II		
Neutron Radiographic Physics Course	14	14
Neutron Radiographic Technique Course	26	26
Total	40	40

*A High school graduate or equivalent.

*B Completion with passing grades of at least two years of engineering or science study at a university, college, or technical school.

Recommended Training for Level I Neutron Radiographic Testing

Neutron Radiographic Equipment Operating and Emergency Instructions Course

Note: It is recommended that the trainee receive instruction in this course prior to performing work in neutron radiography.

1. **Personnel Monitoring**
   a. Personnel-monitoring dosimeters
      (1) Types
      (2) Reading
      (3) Record-keeping
   b. Permissible personnel-exposure limits

2. **Radiation-Survey Instruments**
   a. Types of instruments
   b. Reading and interpreting meter indications
   c. Calibration frequency
   d. Calibration expiration—actions to be taken
   e. Battery check—importance

3. **Radiation-Area Surveys**
   a. Type and quantity of radiation
   b. Posting
      (1) Radiation areas
      (2) High-radiation areas
   c. Establishment of time limits

4. **Radioactivity**
   a. Radioactive components (fuel, sources, etc.)
   b. Induced radioactivity—due to neutron radiography
      (1) Handling of radioactive components
      (2) Decay of radioactive components
      (3) Shipping of radioactive components

5. **Radiation-Area Work Practices—Safety**
   a. Use of time, shielding, and distance to reduce personnel radiation exposure
   b. Restricted areas
   c. Radioactive contamination
      (1) Clothing requirements
      (2) Contamination control
      (3) Contamination cleanup
   d. Specific procedures

6.* **Explosive-Device Safety**
   a. Static electricity
   b. Grounding devices
   c. Clothing requirements
   d. Handling and storage requirements and procedures
   e. Shipping and receiving procedures

7. **State and Federal Regulations**
   a. Nuclear Regulatory Commission (NRC) and agreement states authority
   b. Occupational Safety and Health Administration (OSHA)
   c. Department of Transportation (DOT)
   d.* State and federal explosive-licensing requirements

*Required only by those personnel who will be involved in neutron radiography of explosive devices.

Total recommended hours of instruction for this course:
   Classification A - 8 hours
   Classification B - 8 hours

Basic Neutron Radiographic Physics Course

1. **Introduction**
   a. History of industrial neutron radiography
   b. General principles of examination of materials by penetrating radiation
   c. Relationship of penetrating neutron radiation, radiography, and radiometry
   d. Comparison with other NDT methods, particularly with X-rays and gamma rays
   e. General areas of application
      (1) Imaging
      (2) Metrology
      (3) Product

2. **Physical Principles**
   a. Sources for neutron radiography (general description)
      (1) Isotopes
      (2) Nuclear reactors
      (3) Accelerators
   b. Interaction between neutrons and matter
      (1) Absorption
         (a) Thermal neutrons
         (b) Resonance neutrons
         (c) Fast neutrons
      (2) Scatter
         (a) Elastic
         (b) Inelastic
   c. Neutron radiography techniques
      (1) Film imaging techniques
      (2) Nonfilm imaging techniques
   d. Glossary of terms and units of measure

3. **Radiation Sources for Neutrons (Specific Description)**
   a. Reactors
      (1) Principle of fission chain reactions
      (2) Neutron thermalization (slowing down)
      (3) Thermal neutron flux
   b. Accelerators
      (1) Types of accelerators
      (2) Neutron-producing reactions
   c. Isotopic sources
      (1) Radioisotope + Be
         (a) α − Be
         (b) γ − Be
      (2) Radioisotope + D
         (a) γ − D
      (3) Spontaneous fission
         (a) ^{252}Cf

4. **Personnel Safety and Radiation Protection**
   a. Hazards of excessive exposure
      (1) General—beta-, gamma-radiation
      (2) Specific neutron hazards
         (a) Relative biological effectiveness
         (b) Neutron activation
   b. Methods of controlling radiation dose
      (1) Time
      (2) Distance
      (3) Shielding
   c. Specific equipment requirements
      (1) Neutron monitoring dosimeters
      (2) Gamma-ray monitoring dosimeters
      (3) Radiation survey equipment
         (a) Beta/gamma
         (b) Neutron
      (4) Recording/record-keeping
   d. Radiation work procedures
   e. Federal, state, and local regulations

Total recommended hours of instruction for this course:
    Classification A - 7 hours
    Classification B - 4 hours

Basic Neutron Radiographic Technique Course

1. **Radiation-Detection Imaging**
   a. Converter screens
      (1) Principles of operation
      (2) Direct-imaging screens
      (3) Transfer-imaging screens
   b. Film—principles, properties, and uses with neutron converter screens
      (1) Radiation response
      (2) Vacuum/contact considerations
      (3) Radiographic speed
      (4) Radiographic contrast
   c. Track-etch
      (1) Radiation response
      (2) Vacuum/contact considerations
      (3) Radiographic speed
      (4) Radiographic contrast

2. **Neutron Radiographic Process: Basic Imaging Considerations**
   a. Definition of sensitivity (including penetrameters)
   b. Contrast and definition
      (1) Neutron energy and neutron screen relationship
      (2) Effect of scattering in object
   c. Geometric principles
   d. Generation and control of scatter
   e. Choice of neutron source
   f. Choice of film
   g. Use of exposure curves
   h. Cause of correction of unsatisfactory radiographs
      (1) High film density
      (2) Low film density
      (3) High contrast
      (4) Low contrast
      (5) Poor definition
      (6) Excessive film fog
      (7) Light leaks
      (8) Artifacts
   i. Arithmetic of exposure

3. **Test Result Interpretation**
   a. Relationship between X-ray and n-ray
   b. Effects on measurement and interpretation of test
   c. Administrative control of test quality by interpreter
   d. Familiarization with image

Total recommended hours of instruction for this course:
    Classification A - 13 hours
    Classification B - 8 hours

Recommended Training for
Level II Neutron Radiographic Testing
Neutron Radiographic Physics Course

1. **Introduction**
   a. General principles of examination of materials by penetrating radiation
   b. Relationship of penetrating neutron radiation, radiography, and radiometry
   c. Comparison with other methods, particularly with X-rays and gamma rays
   d. Specific areas of application in industry

2. **Review of Physical Principles**
   a. Nature of penetrating radiation (all types)
      (1) Particles
      (2) Wave properties
      (3) Electomagnetic waves
      (4) Fundamentals of radiation physics
      (5) Sources of radiation
         (a) Electronic sources
         (b) Isotopic sources
         (c) Nuclear reactors
         (d) Accelerators
   b. Interaction between penetrating radiation and matter (neutron and gamma ray)
      (1) Absorption
      (2) Scatter
      (3) Other interactions
   c. Glossary of terms and units of measure

3. **Radiation Sources for Neutrons**
   a. Neutron sources—general
      (1) Reactors
         (a) Principle of fission chain reactions
         (b) Fast-neutron flux—energy and spatial distribution
         (c) Neutron thermalization
         (d) Thermal-neutron flux—energy and spatial distribution
      (2) Accelerators
         (a) Types of accelerators
         (b) Neutron-producing reactions
         (c) Available yields and energy spectra
      (3) Isotopic sources
         (a) Radioisotope + Be
         (b) Radioisotope + D
         (c) Spontaneous fission $-^{252}$Cf
      (4) Beam design
         (a) Source placement
         (b) Collimation
         (c) Filtering
         (d) Shielding

4. **Radiation Detection**
   a. Imaging
      (1) Converter screens
         (a) Principles of operations
         (b) Types of screens
            (1_1) Direct exposure
            (2_1) Transfer exposure
            (3_1) Track-etch process
            (4_1) Spectral sensitivity (each process)
      (2) Film—principles, properties, use with neutron converter screens
         (a) Material examination
         (b) Monitoring
      (3) Fluoroscopy
         (a) Fluorescent screen
         (b) Image amplification
         (c) Cine techniques
      (4) Direct TV viewing
      (5) Special instrumentation associated with above techniques
   b. Nonimaging devices
      (1) Solid-state
         (a) Scintillometers
         (b) Photoresistive devices
         (c) Other
      (2) Gaseous
         (a) Proportional counters
         (b) Geiger counters
         (c) Ionization chambers
         (d) Other
      (3) Neutron detectors
         (a) Boron-based gas counters
         (b) Fission counters
         (c) Helium-3 detectors
         (d) Lithium-based scintillators
         (e) Instrumentation
            (1_1) Rate meters
            (2_1) Counters
            (3_1) Amplifiers and preamplifiers
            (4_1) Recording readouts
            (5_1) Other

5. **Personnel Safety and Radiation Protection**
   a. Hazards of excessive exposure
      (1) General—beta-, gamma ray
      (2) Specific neutron hazards
         (a) Relative biological effectiveness (RBE)
         (b) Neutron activation of components
   b. Methods of controlling accumulated radiation dose
      (1) Time
      (2) Distance
      (3) Shielding
   c. Specific equipment requirements
      (1) Neutron monitoring equipment
      (2) Gamma-ray monitoring equipment
      (3) Survey
      (4) Recording
      (5) Exposure shields and/or rooms
         (a) Operation

(b) Alarms
d. Operation and emergency procedures
e. Federal, state, and local regulations

Total recommended hours of instruction for this course:

Classification A - 14 hours
Classification B - 14 hours

Neutron Radiographic Technique Course

1. **Neutron Radiographic Process**
   a. Basic neutron-imaging considerations
   - (1) Definition of sensitivity (including penetrameters)
   - (2) Contrast and definition
     - (a) Neutron energy and neutron screen relationship
     - (b) Effect of scattering in object
     - (c) Exposure vs. foil thickness
   - (3) Geometric principles
   - (4) Intensifying screens
     - (a) Fluorescent (neutron-sensitive)
     - (b) Metallic (neutron-sensitive)
   - (5) Generation and control of scatter
   - (6) Choice of source
   - (7) Choice of film/detector
   - (8) Use of exposure curves and process by which they are generated
   - (9) Fluoroscopic inspection
     - (a) Theory of operation
     - (b) Applications
     - (c) Limitations
   - (10) Film processing
     - (a) Darkroom procedures
     - (b) Darkroom equipment and chemicals
     - (c) Film processing dos and don'ts
   - (11) Viewing of radiographs
     - (a) Illuminator requirements (intensity)
     - (b) Background lighting
     - (c) Judging quality of neutron radiographs
   - (12) Causes and correction of unsatisfactory radiographs
     - (a) High film density
     - (b) Insufficient film density
     - (c) High contrast
     - (d) Low contrast
     - (e) Poor definition
     - (f) Excessive neutron scatter
     - (g) Fog
     - (h) Light leaks
     - (i) Artifacts
   - (13) Arithmetic of exposure and of other factors affecting neutron radiographs

   b. Miscellaneous applications
   - (1) Blocking and filtering
   - (2) Multifilm techniques
   - (3) Enlargement and projection
   - (4) Stereoradiography
   - (5) Triangulation methods
   - (6) Autoradiography
   - (7) Flash neutron radiography
   - (8) "In-motion" radiography and fluoroscopy
   - (9) Backscatter neutron radiography
   - (10) Neutron tomography
   - (11) Micro-neutron radiography
   - (12) Causes of "diffraction" effects and minimization of interference with test
   - (13) Determination of focal-spot size
   - (14) Panoramic techniques
   - (15) Altering film contrast and density
   - (16) Gaging and control processes

2. **Test Result Interpretation**
   a. Basic factors
   - (1) General aspects (relationship between X-ray and neutron radiographs)
   - (2) Effects on measurement and interpretation of test
   - (3) Administrative control of test quality by interpreter
   - (4) Familiarization with image

   b. Material considerations
   - (1) Metallurgy or other material consideration as it affects use of item and test results
   - (2) Materials-processing effects on use of item and test results
   - (3) Discontinuities—their causes and effects
   - (4) Radiographic appearance of discontinuities

   c. Codes, standards, specifications, and procedures
   - (1) Thermal neutron radiography
   - (2) Resonance neutron radiography
   - (3) Other applicable codes, etc.

Total recommended hours of instruction for this course:

Classification A - 26 hours
Classification B - 26 hours

Recommended Training References

Neutron Radiographic Testing

American Society for Metals. *Nondestructive Inspection and Quality Control: Metals Handbook*, Vol. 11, 8th ed. Metals Park, OH, 1976.*

American Society for Nondestructive Testing. *Question and Answer Book F: Neutron Radiographic Test Method, Levels I, II, and III.* Columbus, OH, 1980.*

American Society for Testing and Materials. "Metallography; Nondestructive Testing," Vol. 03.03. Philadelphia, PA, Latest Edition.*

Atomic Energy Review, Vol. 15, No. 2, June 1977.

Berger, H. *Neutron Radiography.* Amsterdam, Netherlands: Elsevier Publishing Co., 1965.

__________. "Neutron Radiography," *Annual Reviews of Nuclear Science,* Vol. 21. Palo Alto, CA: Annual Review, Inc., 1971.

__________. *Neutron Radiography and Gaging.* Philadelphia, PA: American Society for Testing and Materials, 1967.

Bryant, Lawrence A., and Paul McIntire, eds. *Nondestructive Testing Handbook: Radiography and Radiation Testing*, Vol. 3, 2nd ed. Columbus, OH: American Society for Nondestructive Testing, 1985.*

Herz, R. *The Photographic Action of Ionizing Radiations.* New York: Wiley-Interscience, 1969.

Mix, Paul E. *Introduction to Nondestructive Testing: A Training Guide.* New York: John Wiley & Sons, 1987.*

Neutron Radiography Handbook. Holland and Boston: D. Reidel Publishing Co., 1981.

Radiographic Testing, Classroom Training Handbook (CT-6-6). San Diego, CA: General Dynamics/Convair Division, 1967.*

Radiographic Testing, Programmed Instruction Handbook (PI-4-6). San Diego, CA: General Dynamics/Convair Division, 1983.*

Radiography in Modern Industry, 4th ed. Rochester, NY: Eastman Kodak Co., 1980.*

Sensitometric Properties of X-Ray Films. Rochester, NY: Eastman Kodak Co., 1974.

Standards for Protection against Radiation, Title 10: Rules and Regulations, Part 20. Washington, DC: U.S. Government Printing Office.

Tyufakov, N. D., and A. S. Shtan. *Principles of Neutron Radiography*, TT76-52048. New Delhi, India: Amerind Publishing Co., 1979.

*Available from the American Society for Nondestructive Testing, Columbus, OH.

Leak Testing Method

(LT — Training Course Outline — TC-7)

	Recommended Hours of Instruction	
	A*	B*
Level I		
Fundamentals in Leak Testing Course	14	8
Safety in Leak Testing	14	8
Leak Testing Methods	<u>14</u>	<u>8</u>
Total	42	24
Level II		
Principles of Leak Testing Course	24	12
Pressure and Vacuum Technology Course	12	6
Leak Test Selection Course	<u>12</u>	<u>6</u>
Total	48	24

*A High school graduate or equivalent.

*B Completion with passing grades of at least two years of engineering or science study at a university, college, or technical school.

Hours of instruction may vary according to the technique being instructed, i. e., the hours of instruction for one or two techniques may be less than for four techniques. **The minimum recommended training hours are shown in Table 6.3.1, page 7.**

Recommended Training for Level I Leak Testing

Fundamentals in Leak Testing Course

1. **Introduction**
   a. History of leak testing
   b. Reasons for leak testing
      (1) Material loss prevention
      (2) Contamination
      (3) Component/system reliability
      (4) Pressure-differential maintenance
      (5) Personnel/public safety
   c. Functions of leak testing
      (1) Categories
      (2) Applications
   d. Training and certification

2. **Leak Testing Fundamentals**
   a. Terminology
      (1) Leakage terms
      (2) Leakage tightness
      (3) Quantitative/semiquantitative
      (4) Sensitivity/calibration terms
   b. Leak testing units
      (1) Mathematics in leak testing
      (2) Exponential notation
      (3) Basic and fundamental units
      (4) Système Internationale (SI) units
   c. Physical units in leak testing
      (1) Volume and pressure
      (2) Time and temperature
      (3) Absolute values
      (4) Standard or atmospheric conditions
      (5) Leakage measurement
   d. Leak testing standards
      (1) Capillary or permeation
      (2) National Institute of Standards and Technology (NIST) standards
      (3) System vs. instrument calibration
      (4) Inaccuracy of calibration
   e. Flow characteristics
      (1) Gas flow
      (2) Liquid flow
      (3) Correlation of leakage rates
      (4) Anomalous leaks
      (5) Leak clogging
   f. Vacuum fundamentals
      (1) Introduction to vacuum
         (a) Terminology
         (b) Principles
         (c) Units of pressure
      (2) Characteristics of gases
         (a) Kinetic theory
         (b) Mean free path
         (3) Gas laws
         (4) Quantity, throughput, and conductance of gas
            (a) Quantity
               (1_1) Comparison with an electric circuit
               (2_1) Comparison with water flow
            (b) Conductance analogy with electrical resistance
               (1_1) Resistance connected in series
               (2_1) Resistance connected in parallel
   g. Vacuum system operation
      (1) Effects of evacuating a vessel
      (2) Pump-down time
   h. Vacuum system characteristics
      (1) General
         (a) Operating limits
         (b) Rate of pressure rise — measurement
      (2) Vacuum pumps
         (a) Mechanical pumps (positive displacement)
            (1_1) Oil-sealed rotary pumps
               (a_1) Construction
               (b_1) Operation
               (c_1) Pump fluids
               (d_1) Difficulties with rotary pumps
               (e_1) Care of rotary pumps
            (2_1) Mechanical booster pumps
         (b) Vapor (diffusion) pumps
            (1_1) Construction
            (2_1) Operation
            (3_1) Pump fluids
            (4_1) Difficulties with diffusion pumps
            (5_1) Diagnosis of diffusion pump trouble
         (c) Sublimation pumps (getter pumps)
         (d) Ion pumps
         (e) Turbomolecular pumps
         (f) Absorption pumps
         (g) Cryopumps

Total recommended hours of instruction for this course:
   Classification A - 14 hours
   Classification B - 8 hours

Safety in Leak Testing Course

Note: It is recommended that the trainee, as well as all other leak testing personnel, receive instruction in this course prior to performing work in leak testing.

1. **Safety Considerations**
   a. Personnel and the public
   b. Product serviceability
   c. Test validity
   d. Safe work practices

2. **Safety Precautions**
   a. Explosive/implosive hazards
   b. Flammability, ignitibility, combustibility hazards
   c. Toxicity and asphyxiation hazards
   d. Cleaning and electrical hazards

3. **Pressure Precautions**
   a. Pressure test vs. proof test
   b. Preliminary leak testing
   c. Pressurization check
   d. Design limitations
   e. Equipment and setup

4. **Safety Devices**
   a. Pressure control valves and regulators
   b. Pressure relief valves and vents
   c. Flow rate of regulator and relief valves

5. **Hazardous and Tracer Gas Safety**
   a. Combustible gas detection and safety
   b. Toxic gas detection and safety
   c. Oxygen-deficiency detectors
   d. Radioisotope detection

6. **Types of Monitoring Equipment**
   a. Area monitors
   b. Personnel monitors
   c. Leak-locating devices

7. **Safety Regulations**
   a. State and federal regulations
   b. Safety codes/standards
   c. Hazardous gas standards
   d. Nuclear Regulatory Commission (NRC) radiation requirements

Total recommended hours of instruction for this course:
   Classification A - 14 hours
   Classification B - 8 hours

Leak Testing Methods Course

1. **The following leak testing methods may be incorporated as applicable.**
   a. Each of these methods can be further divided into major techniques as shown in the following examples.
      (1) Bubble testing
         (a) Immersion
         (b) Film solution
      (2) Ultrasonic testing
         (a) Sonic/mechanical flow
         (b) Sound generator
      (3) Voltage discharge testing
         (a) Voltage spark
         (b) Color change
      (4) Pressure leak testing
         (a) Hydrostatic
         (b) Pneumatic
      (5) Ionization
         (a) Photoionization
         (b) Flame ionization
      (6) Conductivity
         (a) Thermal conductivity
         (b) Solid state
      (7) Radiation absorption
         (a) Infrared
         (b) Ultraviolet
         (c) Laser
      (8) Chemical-based
         (a) Chemical penetrants
         (b) Chemical gas tracer (colorimetric)
      (9) Halogen detector
         (a) Halide torch
         (b) Electron capture
         (c) Halogen diode
      (10) Pressure change measurement
         (a) Absolute
         (b) Reference
         (c) Pressure rise
         (d) Flow
         (e) Pressure decay
         (f) Volumetric
      (11) Mass spectrometer
         (a) Helium or argon leak detector
         (b) Residual gas analyzer
      (12) Radioisotope

2. **Leak Testing Method Course Outline**
   a. The following may be applied to any of the listed methods.
   b. Terminology
   c. Basic techniques and/or units
      (1) Leak location—measurement/monitoring
      (2) Visual and other sensing devices
      (3) Various techniques
   d. Testing materials and equipment
      (1) Materials, gases/fluids used
      (2) Control devices and operation
      (3) Instrument/gages used
      (4) Range and calibration of instrument/gages
   e. Testing principles and practices
      (1) Pressure/vacuum and control used

(2) Principles of techniques used
(3) Effects of temperature and other atmospheric conditions
(4) Calibration for testing
(5) Probing/scanning or measurement/ monitoring
(6) Leak interpretation evaluation
f. Acceptance and rejection criteria

g. Safety concerns
h. Advantages and limitations
i. Codes/standards

Total recommended hours of instruction for this course:
Classification A - 14 hours
Classification B - 8 hours

Recommended Training for Level II Leak Testing

Principles of Leak Testing Course

1. **Introduction**
   a. Leak testing fundamentals
      (1) Reasons for leak testing
      (2) Functions of leak testing
      (3) Terminology
      (4) Leak testing units
      (5) Leak conductance
   b. Leak testing standards
      (1) Leak standards
      (2) National Institute of Standards and Technology (NIST) traceability and calibration
      (3) Instrument calibration vs. test qualification
      (4) System calibration techniques
      (5) Inaccuracy of calibration
      (6) Tracer-gas leak rate/air-equivalent leak rate
   c. Leak testing safety
      (1) Safety considerations
      (2) Safety precautions
      (3) Pressure precautions
      (4) Tracer gas safety and monitoring
      (5) Safety devices
      (6) Cleaning and electrical hazards
      (7) Safe work practices
      (8) Safety regulations
   d. Leak testing procedure
      (1) Basic categories and techniques
      (2) Leak location vs. leakage measurement
      (3) Pressurization or evacuation
      (4) Sealed units with or without tracer gas
      (5) Units accessible from one or both sides
      (6) System at, above, or below atmospheric pressure
   e. Leak testing specifications
      (1) Design vs. working conditions
      (2) Pressure and temperature control
      (3) Types of leak testing methods
      (4) Sensitivity of leak testing methods
      (5) Test method and sensitivity needed
      (6) Preparation of a leak testing specification
   f. Detector/instrument performance factors
      (1) Design and use
      (2) Accuracy and precision
      (3) Linearity (straight/logarithmic scale)
      (4) Calibration and frequency
      (5) Response and recovery time

2. **Physical Principles in Leak Testing**
   a. Physical quantities
      (1) Fundamental units
      (2) Volume and pressure
      (3) Time and temperature
      (4) Absolute values
      (5) Standard vs. atmospheric conditions
      (6) Leakage rates
   b. Structure of matter
      (1) Atomic theory
      (2) Ionization and ion pairs
      (3) States of matter
      (4) Molecular structure
      (5) Diatomic and monatomic molecules
      (6) Molecular weight
   c. Gas principles and laws
      (1) Brownian movement
      (2) Mean free path
      (3) Pressure and temperature effects on gases
      (4) Pascal's law of pressure
      (5) Charles' and Boyle's gas laws
      (6) Ideal gas law
      (7) Dalton's law of partial pressure
      (8) Vapor pressure and effects in vacuum
   d. Gas properties
      (1) Kinetic theory of gases
      (2) Graham law of diffusion
      (3) Stratification
      (4) Avogadro's principle
      (5) Gas law relationship
      (6) General ideal gas law
      (7) Gas mixture and concentration
      (8) Gas velocity, density, and viscosity

3. **Principles of Gas Flow**
   a. Standard leaks
      (1) Capillary
      (2) Permeation
   b. Modes of gas flow
      (1) Molecular and viscous
      (2) Transitional
      (3) Laminar, turbulent, sonic
   c. Factors affecting gas flow
   d. Geometry of leakage path
      (1) Mean free flow of fluid
      (2) Clogging and check valve effects
      (3) Irregular aperture size
      (4) Leak rate vs. cross section of flow
      (5) Temperature and atmospheric conditions
      (6) Velocity gradient vs. viscosity
      (7) Reynolds number vs. Knudsen number

Total recommended hours of instruction for this course:

Classification A - 24 hours

Classification B - 12 hours

Pressure and Vacuum Technology Course

1. **Pressure Technology**
   a. Properties of a fluid
      (1) What is a fluid?
      (2) Liquid vs. gas
      (3) Compressibility
      (4) Partial and vapor pressure
      (5) Critical pressure and temperature
      (6) Viscosity of a liquid
      (7) Surface tension and capillarity of a liquid
   b. Gas properties
      (1) Review of gas properties
      (2) What is a perfect/ideal gas?
      (3) Pressure and temperature effects on gases
      (4) Viscosity of a gas
      (5) Gas flow modes
      (6) Gas flow conductance
      (7) Dynamic flow measurements
      (8) Factors affecting gas flow
   c. Pressurization
      (1) Pressure measurements
      (2) Types of pressure gages
         (a) Bourdon or diaphragm
         (b) Manometers
      (3) Pressure control and procedure
      (4) Mixing of gases
      (5) Tracer gases and concentration
      (6) Pressure hold time
      (7) Pressure vs. sensitivity
      (8) Gage calibration
         (a) Working range
         (b) Frequency
         (c) Master gage vs. dead-weight tester
   d. Leak testing background/noise variables
      (1) Atmospheric changes
      (2) Liquid/air temperature correction
      (3) Vapor pressure (evaporation/condensation)
      (4) Vapor/moisture pockets
      (5) Geometry/volume changes
      (6) Surface/internal vibration waves
   e. Detector/instrument performance variables
      (1) Instrument calibration variables
      (2) Limits of accuracy
      (3) Intrinsic and inherent safety performance
      (4) Protection for electromagnetic interference, radio frequency interference, shock, etc.
      (5) Flooding, poisoning, contamination
   f. Measurement and data documentation
      (1) Experimental, simulation, and/or preliminary testing
      (2) Analysis of background/noise variables
      (3) Analysis of leakage indications/signals
      (4) Validation and error analysis
      (5) Interpretation and evaluation of results
      (6) Documentation of data and test results

2. **Vacuum Technology**
   a. Nature of vacuum
      (1) What is a vacuum?
      (2) Vacuum terminology
      (3) Degrees of vacuum
      (4) Mean free path in a vacuum
      (5) Gas flow in a vacuum
   b. Vacuum measurement
      (1) Pressure units in a vacuum
      (2) Absolute vs. gage pressure
      (3) Mechanical gages
         (a) Bourdon or diaphragm
         (b) Manometer (U-tube or McLeod)
         (c) Capacitance manometer
      (4) Electrical gages
         (a) Thermal conductivity
         (b) Ionization
      (5) Gage calibration—full range
   c. Vacuum pumps
      (1) Types of vacuum pumps
      (2) Mechanical pumps
         (a) Reciprocating vs. rotary
         (b) Roots, turbomolecular, drag pumps
      (3) Nonmechanical pumps
         (a) Fluid entrainment or diffusion
         (b) Condensation or sorption
      (4) Pump oils
      (5) Pumping speed and pump-down time
   d. Vacuum materials
      (1) Outgassing—vapor pressure
      (2) Elastomers, gaskets, O-rings
      (3) Metals, metal alloys, and nonmetals
         (a) Carbon steel vs. stainless steel
         (b) Aluminum, copper, nickel, and alloys
      (4) Nonmetals
         (a) Glass, ceramics
         (b) Plastics, Tygon™, etc.
      (5) Joint design
         (a) Sealed joint
         (b) Welded/brazed joint
         (c) Mechanical joint
      (6) Vacuum greases and sealing materials
      (7) Tracer gas permeation through materials
   e. Design of a vacuum system
      (1) Production of a vacuum
         (a) Removal of gas molecules
         (b) Gas quantity or throughput
         (c) Conductance
      (2) Stages of vacuum pumping
         (a) Various vacuum pumps
         (b) Various traps and baffles
         (c) Pumping stages or sequences
      (3) Vacuum valve location
         (a) Vacuum valve design and seat leakage
         (b) Isolation and protection
         (c) Automatic vs. manual

(d) Venting
f. Maintenance and cleanliness
    (1) Maintenance of vacuum equipment
        (a) Under constant vacuum
        (b) Dry gas (nitrogen)
    (2) Routing oil changes
    (3) System cleanliness
        (a) Initial cleanliness
        (b) Cleaning procedures and effects
            on leak location and measurement
        (c) Continued cleanliness
g. Analysis and documentation

(1) Analysis of outgassing and background
    contamination
(2) Instrument/system calibration
(3) Analysis of leakage indications/signals
(4) Interpretation and evaluation
(5) Documentation of calibration and test
    results

Total recommended hours of instruction for this
course:
    Classification A - 12 hours
    Classification B - 6 hours

Leak Test Selection Course

1. Choice of Leak Testing Procedure
a. Basic categories of leak testing
    (1) Leak location
    (2) Leakage measurement
    (3) Leakage monitoring
b. Types of leak testing methods
    (1) Specifications
    (2) Sensitivity
c. Basic techniques
    (1) Pressurization or evacuation

(2) Sealed unit with or without tracer
    gases
(3) Probing or visual leak location
(4) Tracer or detector probing
(5) Accumulation techniques

Total recommended hours of instruction for this
course:
    Classification A - 12 hours
    Classification B - 6 hours

Recommended Training References

Leak Testing Method

American National Standards Institute. "Containment System Leakage Testing Requirements (ANSI/ANS 56.8)." New York, 1981.

__________. "Leakage-rate Testing of Containment Structures for Nuclear Reactors (N45.4)." New York, 1972.

__________. "Testing of Nuclear Air-Cleaning Systems (ANSI/ASME N510)." Néw York.

American Society of Mechanical Engineers. "Water and Steam in the Power Cycle (Purity and Quality, Leak Detection and Measurement) (PTC 19.11, Part II)." New York, 1970.

__________. "ASME Boiler and Pressure Vessel Inspection Code—Section V, Article 10, Leak Testing." New York.

American Society for Metals. *Nondestructive Inspection and Quality Control: Metals Handbook*, Vol. 11, 8th ed. Metals Park, OH, 1976.*

American Society for Nondestructive Testing. *Question and Answer Book HB: Bubble Leak Testing, Levels I, II, and III*. Columbus, OH, 1985.*

__________. *Question and Answer Book HH: Halogen Diode Detector Leak Testing, Levels I, II, and III*. Columbus, OH, 1985.*

__________. *Question and Answer Book HM: Mass Spectrometer Test Method, Levels I, II, and III*. Columbus, OH, 1985.*

__________. *Question and Answer Book HP: Pressure Change Measurement Testing, Levels I, II, and III*. Columbus, OH, 1985.*

American Society for Testing and Materials. "Metallography; Nondestructive Testing," Vol. 03.03. Philadelphia, PA, Latest Edition.*

American Vacuum Society. "Calibration of Leak Detectors of the Mass Spectrometer Type (2.1)." New York, 1973.

__________. "Method for Vacuum Leak Detection (2.2)." New York, 1968.

__________. "Procedure for the Calibration of Gas Analyzers of the Mass Spectrometry Type (2.3)." New York, 1973.

Drinkwine and Lichlman. *Partial Pressure Analyzer and Analysis*. New York: American Vacuum Society, 1979.

Dushman, S. "Scientific Foundation of Vacuum Technique," Third Printing. Somerset, NJ: John Wiley & Sons, Inc., 1965.

Guthrie, A. "Vacuum Technology." Somerset, NJ: John Wiley & Sons, Inc., 1963.

Halmshaw, R., ed. *Mathematics and Formulas in NDT*. British Institute of Non-Destructive Testing, 1978.*

Instrument Society of America. "Light Water Reactor Coolant Pressure Boundary Leak Detection (S67.03)." Pittsburgh, PA.

International Organization for Standardization. "Vacuum Technology—Mass Spectrometer Type Leak Detector (3530)." Geneva, Switzerland.

"Introduction to Vacuum and Leak Detection." Plainview, NY: Veeco Instruments, Inc., 1980.

"Leak Detection Manual (Halogen), 1D-4816." Shenandoah, GA: General Electric.

"Leak Detection Manual (Helium), 4820-48000." Wilmington, DE: CEC/Du Pont, 1982.

"Leakage Testing Handbook" by General Electric and NASA under contracts CR-952 and NAS7-396, NASA Report #N69-38843, available as report IST-295 from National Technical Information Services, Springfield, VA.

McGonnagle, Warren J. *Nondestructive Testing*, 2nd ed. New York: Gordon and Breach, 1975.

McMaster Robert C., ed. *Nondestructive Testing Handbook: Leak Testing*, Vol. 1, 2nd ed. Columbus, OH: American Society for Nondestructive Testing, 1982.*

Military Publications**

"Leak Detection Compound, Oxygen Systems (MIL-I-2556C)."

"Leak Detector, Refrigerant Gas; Acetylene Burning with Search Hose (MIL-L-3516C)."

"Leak Detector, Full System (MIL-L-83774)."

"Liquid Dye for Leak Detection (MIL-D-81298)."

Mix, Paul E. *Introduction to Nondestructive Testing: A Training Guide*. New York: John Wiley & Sons, Inc., 1987.*

Modey and Brown, eds. *History of Vacuum Science and Technology*. New York: American Vacuum Society, 1984.

"Nondestructive Testing—A Survey." NASA Report SP-5113, Southwest Research Institute, available as Report N73-28517 from National Technical Information Service, Springfield, VA.

O'Hanlon, J. *A User's Guide to Vacuum Technology*. New York: John Wiley & Sons, Inc.

Quality Assurance Provision for Nondestructive Testing Handbook, QE-702-14E. Alabama: U.S. Army Redstone Arsenal.*

Tompkin, H. *Introduction to the Fundamentals of Vacuum Technology*. New York: American Vacuum Society, 1984.

U.S. Government, Department of Energy. "Nondestructive Evaluation Criteria for Use of ASME Section III and USASI B31.7. Div., Reactor Dev. and Tech."**

__________. "Primary Reactor Containment Leakage Testing for Water-Cooled Power Reactors (Appendix J Title 10 CFR Part 50)."**

U.S. Government Fed. Test Method Standards**

"Leak Testing (Helium Mass Spectrometer) (#151 b-method 441)."

"Leak Testing (Pressurized Gas) (#151 b-method 442)."

"Leak Testing (Vacuum) (#151 b-method 443)."

"Vacuum Technology: Its Foundation, Formula, and Tables." E. Syracuse, NY: Inficon Leybold-Heraeus, 1980.

Varian Assn. "Introduction to Helium Mass Spectrometer Leak Detection." Lexington, MA, 1980.

Weast, R. C. *Handbook of Chemistry and Physics.* Boca Raton, FL: CRC Press.

Wilson, N., and L. Beavis. *Handbook of Vacuum Leak Detection.* New York: American Vacuum Society, 1979.

*Available from the American Society for Nondestructive Testing, Columbus, OH.

**Available from the Naval Publications and Forms Center, 5801 Tabor Ave., Philadelphia, PA 19120.

Acoustic Emission Testing Method
(AET — Training Course Outline — TC-8)

	Recommended Hours of Instruction	
	A*	B*
Level I		
Basic Acoustic Emission Physics Course	12	10
Basic Acoustic Emission Technique Course	<u>28</u>	<u>22</u>
Total	40	32
Level II		
Acoustic Emission Physics Course	6	6
Acoustic Emission Technique Course	<u>34</u>	<u>34</u>
Total	40	40

***A** High school graduate or equivalent.

***B** Completion with passing grades of at least two years of engineering or science study at a university, college, or technical school.

This is a progressive training course; i.e., consideration as Level I is based on satisfactory completion of the Level I training course; consideration as Level II is based on satisfactory completion of both Level I and Level II training courses.

Topics in the training outline may be deleted or expanded to meet the employer's specific applications or for limited certification and may be accompanied by a corresponding change in training hours.

Recommended Training for
Level I Acoustic Emission Testing

Basic Acoustic Emission Physics Course

1. **Principles of Acoustic Emission Testing**
    a. Characteristics of acoustic emission
        (1) Continuous emission
        (2) Burst emission
        (3) Emission levels and frequencies
    b. Sources of acoustic emission
        (1) Sources in crystalline materials—introduction
        (2) Sources in nonmetals—introduction
        (3) Sources in composites—introduction
        (4) Other sources
            (a) Pressure leaks
            (b) Oxide and scale cracking
            (c) Slag cracking
            (d) Frictional sources
            (e) Liquefaction and solidification
            (f) Loose parts, intermittent contact
            (g) Fluids and nonsolids
            (h) Crack closure
    c. Wave propagation—introduction
        (1) Attenuation
        (2) Wave velocity in materials
        (3) Source input vs. signal output
    d. Kaiser effect
        (1) In metals
        (2) In composites
        (3) In other materials
    e. Terminology (refer to AE Glossary, ASTM E610-77)

2. **Detection of Acoustic Emission**
    a. Sensors
        (1) Principles of operation
        (2) Construction
    b. Sensor attachment
        (1) Coupling materials
        (2) Attachment devices
    c. Sensor—preamplifier transmission cables
        (1) Cable types
        (2) Noise susceptibility
        (3) Impedance matching
        (4) Connectors
        (5) Integral electronic sensors

Total recommended hours of instruction for this course:
Classification A - 12 hours
Classification B - 10 hours

Basic Acoustic Emission Technique Course

1. **Instrumentation and Signal Processing**
    a. Signal conditioning
        (1) Preamplifiers
        (2) Amplifiers
        (3) Filters
        (4) Units of gain—measurement
    b. Signal processing
        (1) Waveform characteristics
        (2) Discrimination techniques
        (3) Distribution techniques
        (4) Location techniques
    c. Data display and recording techniques
        (1) Audio indicators
        (2) X-Y and strip-chart recording
        (3) Digital counters and processors
        (4) Oscilloscopes
        (5) Magnetic recorders
        (6) Others
    d. Acoustic emission test systems
        (1) Single-channel systems
        (2) Multichannel systems
        (3) Dedicated industrial systems

2. **Acoustic Emission Test Procedures**
    a. Calibration of test equipment
        (1) Signal generation techniques
        (2) Adjustment of equipment controls
        (3) Discrimination technique adjustments
        (4) Recognition of severe attenuation near failure
    b. Data display and interpretation
        (1) Selection and display mode
        (2) Comparison with calibration signals
        (3) Source evaluation by other NDT methods
    c. Noise identification
        (1) Electromagnetic noise
        (2) Mechanical noise
    d. Noise discrimination
        (1) Electrical shielding
        (2) Electronic techniques
        (3) Attenuating materials and applications
        (4) Pseudosignal identification and control
    e. Reports
        (1) Purpose
        (2) Content and structure

3. **Applications of Acoustic Emission Testing**
    a. Fundamental studies—introduction
    b. Source detection and location
    c. Other applications
        (1) Wood and wooden structures
        (2) Rock mechanics
        (3) Civil structures

     (a) Bridges
     (b) Buildings
     (c) Dams
     (d) Buried pipelines
     (e) Slopes
     (f) Cavities
     (g) Mines
(4) Leak detection
(5) Composite materials and structures
(6) Refractory structures

     (7) Machinery analysis
     (8) Microseismology
     (9) Acoustic signature analysis
d. Conditions that cause poor acoustic emission applicability

Total recommended hours of instruction for this course:
    Classification A - 28 hours
    Classification B - 22 hours

Recommended Training for
Level II Acoustic Emission Testing

Acoustic Emission Physics Course

1. **Principles of Acoustic Emission Testing**
   a. Characteristics of acoustic emission
      (1) Continuous emission
      (2) Burst emission
      (3) Emission levels and frequencies
   b. Sources of acoustic emission
      (1) Sources in crystalline materials
         (a) Dislocations—plastic deformation
         (b) Phase transformations
         (c) Deformation twinning
         (d) Nonmetallic inclusions
         (e) Subcritical crack growth
            (1_1) Subcritical crack growth under increasing load
            (2_1) Ductile tearing under increasing load
            (3_1) Fatigue crack nucleation and growth
            (4_1) Hydrogen embrittlement cracking
            (5_1) Stress corrosion cracking
      (2) Sources in nonmetals
         (a) Microcracking
         (b) Gross cracking
         (c) Crazing
         (d) Other sources in nonmetals
      (3) Sources in composites
         (a) Fiber fracture
         (b) Matrix cracking
         (c) Delamination, disbonding
      (4) Other sources
         (a) Pressure leaks
         (b) Oxide and scale cracking
         (c) Slag cracking
         (d) Frictional sources
         (e) Liquefaction and solidification
         (f) Loose parts, intermittent contact
         (g) Fluids and nonsolids
         (h) Crack closure
   c. Wave propagation
      (1) Modes of propagation
      (2) Mode conversion, reflection, and refraction
      (3) Attenuation
      (4) Wave velocity in materials
      (5) Specimen geometry effects
   d. Kaiser effect
      (1) In metals
      (2) In composites
      (3) In other materials
   e. Terminology (refer to AE Glossary, ASTM E610-77)

2. **Detection of Acoustic Emission**
   a. Detection processes—piezoelectricity, etc.
   b. Sensors
      (1) Construction
      (2) Conversion efficiencies
      (3) Calibration—sensitivity curve
   c. Sensor attachment
      (1) Coupling materials
      (2) Attachment devices
      (3) Waveguides
   d. Sensor/preamplifier transmission cables
      (1) Cable types
      (2) Noise susceptibility
      (3) Impedance matching
      (4) Connectors
      (5) Integral electronic sensors

Total recommended hours of instruction for this course:
Classification A - 6 hours
Classification B - 6 hours

Acoustic Emission Technique Course

1. **Instrumentation and Signal Processing**
   a. Signal conditioning
      (1) Preamplifiers
      (2) Amplifiers
      (3) Filters
      (4) Units of gain—measurement
   b. Signal processing
      (1) Waveform characteristics
         (a) Amplitude analysis
         (b) Pulse duration analysis
         (c) Rise time analysis
         (d) Event and event-rate processing
      (2) Discrimination techniques
      (3) Distribution techniques
      (4) Location techniques
   c. Data display and recording techniques
      (1) Audio indicators
      (2) X-Y and strip-chart recording
      (3) Digital counters and processors
      (4) Oscilloscopes
      (5) Magnetic recorders
      (6) Other techniques
   d. Acoustic emission test systems
      (1) Single-channel systems
      (2) Multichannel systems
      (3) Dedicated industrial systems

2. **Acoustic Emission Test Procedures**
   a. Factors affecting test equipment selection
      (1) Material being monitored
      (2) Location and nature of emission
      (3) Type of information desired
      (4) Size and shape of test part
   b. Calibration of test equipment
      (1) Signal generation techniques
      (2) Adjustment of equipment controls
      (3) Discrimination technique adjustments
      (4) Recognition of severe attenuation near failure
   c. Special test procedures
      (1) High-temperature/low-temperature tests
      (2) Interrupted tests (including cyclic fatigue)
      (3) Long-term tests
      (4) Tests in high-noise environments
   d. Data display and interpretation
      (1) Selection of display mode
      (2) Emission-source correlation
      (3) Comparison with calibration signals
      (4) Source evaluation by other NDT methods
   e. Noise identification
      (1) Electromagnetic noise
      (2) Mechanical noise
   f. Noise discrimination
      (1) Electrical shielding
      (2) Electronic techniques
      (3) Attenuating materials and applications
      (4) Pseudosignal identification and control
   g. Reports
      (1) Purpose
      (2) Content and structure

3. **Applications of Acoustic Emission Testing**
   a. Fundamental studies
      (1) Fracture mechanics
      (2) Stress corrosion
      (3) Hydrogen embrittlement
      (4) Dislocation movement
      (5) Phase transformation
      (6) Phase stability
      (7) Creep
      (8) Residual stress
      (9) Corrosion
   b. Source detection and location
      (1) Aerospace vehicles
      (2) Pressure vessels and piping
      (3) Nuclear reactors
      (4) Weld monitoring
      (5) Trusses and beams
   c. Other applications
      (1) Wood and wooden structures
      (2) Rock mechanics
      (3) Civil structures
          (a) Bridges
          (b) Buildings
          (c) Dams
          (d) Buried pipelines
          (e) Slopes
          (f) Cavities
          (g) Mines
      (4) Leak detection
      (5) Composite materials and structures
      (6) Refractory structures
      (7) Machinery analysis
      (8) Microseismology
      (9) Acoustic signature analysis
   d. Conditions that cause poor acoustic emission applicability

Total recommended hours of instruction for this course:
   Classification A - 34 hours
   Classification B - 34 hours

Recommended Training References
Acoustic Emission Testing Method

American Society for Metals. *Nondestructive Inspection and Quality Control: Metals Handbook*, Vol 11, 8th ed. Metals Park, OH, 1976.*

American Society for Testing and Materials. "Metallography; Nondestructive Testing," Vol. 03.03. Philadelphia, PA, Latest Edition.*

Drouillard, T. F. *Acoustic Emission: A Bibliography with Abstracts*. New York: Plenum Press, 1978.

Liptai, R. G., D. O. Harris, and C. A. Tatro, eds. *Acoustic Emission—STP 505*. Philadelphia, PA: American Society for Testing and Materials, 1972.

Miller, Ronnie K., and Paul McIntire, eds. *Nondestructive Testing Handbook: Acoustic Emission Testing*, Vol. 5, 2nd ed. Columbus, OH: American Society for Nondestructive Testing, 1987.*

Nichols, R. W., ed. *Acoustic Emission*. Essex, England: Applied Science Publishers, 1976.

Spanner, J. C. *Acoustic Emission: Techniques and Applications*. Evanston, IL: Intex Publishing Co., 1974.*

Spanner, J. C., and J. W. McElroy, eds. *Monitoring Structural Integrity by Acoustic Emission—STP 571*. Philadelphia, PA: American Society for Testing and Materials, 1975.

*Available from the American Society for Nondestructive Testing, Columbus, OH.

Visual Testing Method

(VT — Training Course Outline — TC-9)

	Recommended Hours of Instruction	
	A*	B*
Level I	8	4
Level II	16	8

*A High school graduate or equivalent.

*B Completion with passing grades of at least two years of engineering or science study at a university, college, or technical school.

This is a progressive training course; i.e., consideration as Level I is based on satisfactory completion of the Level I training course; consideration as Level II is based on satisfactory completion of both Level I and Level II training courses.

Topics in the training outline may be deleted or expanded to meet the employer's specific applications or for limited certification and may be accompanied by a corresponding change in training hours.

Recommended Training for Level I Visual Testing

Note: The guidelines listed below should be implemented using equipment and procedures relevant to the employer's industry. No times are given for a specific subject; this should be specified in the employer's written practice. Based upon the employer's product, not all of the referenced subcategories need apply.

1. **Introduction**
   a. Definition of visual testing
   b. History of visual testing
   c. Overview of visual testing applications

2. **Definitions**
   Standard terms and their meanings in the employer's industry

3. **Fundamentals**
   a. Vision
   b. Lighting
   c. Material attributes
   d. Environmental factors
   e. Visual perception
   f. Direct and indirect methods

4. **Equipment (as applicable)**
   a. Mirrors
   b. Magnifiers
   c. Borescopes
   d. Fiberscopes
   e. Closed-circuit television
   f. Light sources and special lighting
   g. Gages, templates, scales, special tools, etc.
   h. Automated systems
   i. Computer-enhanced systems

5. **Employer-Defined Applications**
   (Includes a description of inherent discontinuities)
   a. Mineral-based material
   b. Metallic materials, including welds
   c. Organic-based materials
   d. Other materials (employer-defined)

6. **Visual Testing to Specific Procedures**
   a. Selection of parameters
      (1) Inspection objectives
      (2) Inspection checkpoints
      (3) Sampling plans
      (4) Inspection patterns
      (5) Documented procedures
   b. Test standards/calibration
   c. Classification of indications per acceptance criteria
   d. Reports and documentation

Total recommended hours of instruction for this course:
   Classification A - 8 hours
   Classification B - 4 hours

Recommended Training for Level II Visual Testing

The guidelines listed below should be implemented using equipment and procedures relevant to the employer's industry. The employer should tailor the program to the company's particular application area. Discontinuity cause, appearance, and how to best visually detect and identify these discontinuities should be emphasized. No times are given for a specific subject; this should be specified in the employer's written practice. Depending upon the employer's product, not all the referenced subcategories need apply.

1. **Review of Level I**
   a. Definitions
   b. Fundamentals of visual testing
   c. Equipment
   d. Applications

2. **Vision**
   a. The eye
   b. Vision limitations
   c. Disorders
   d. Employer's vision examination methods

3. **Lighting**
   a. Fundamentals of light
   b. Lighting measurements
   c. Recommended lighting levels
   d. Lighting techniques for inspection

4. **Material attributes**
   a. Cleanliness
   b. Color
   c. Condition
   d. Shapes
   e. Size
   f. Temperature
   g. Texture
   h. Type

5. **Environmental and Physiological Factors**
   a. Atmosphere
   b. Cleanliness
   c. Comfort
   d. Distance
   e. Elevation
   f. Fatigue
   g. Health
   h. Humidity
   i. Mental attitude
   j. Position
   k. Safety
   l. Temperature

6. **Visual Perception**
   a. What your eyes see
   b. What your mind sees
   c. What others perceive
   d. What the designer, engineer, etc. wants you to see

7. **Equipment**
   a. Automated systems
   b. Borescopes
   c. Closed-circuit television
   d. Computer-based systems
   e. Fiberscopes
   f. Gages, templates, scales, etc.
   g. Imaging systems
   h. Light sources and special lighting
   i. Magnifiers
   j. Mirrors
   k. Special optical systems
   l. Standard lighting

8. **Employer-Defined Applications**
   a. Mineral-based material
   b. Metallic materials (including welds)
   c. Organic-based materials
   d. Other materials and products (employer-defined)

9. **Acceptance/Rejection Criteria**
   a. Subjective basis (qualitative)
   b. Objective basis (quantitative)
   c. Evaluation of results per acceptance criteria

10. **Recording and Reports**
    a. Subjective method
    b. Objective method
    c. Recording methods

Total recommended hours of instruction for this course:
Classification A - 16 hours
Classification B - 8 hours

Recommended Training References

Visual Testing Method

American Society for Metals. *Nondestructive Inspection and Quality Control: Metals Handbook,* Vol. 11, 8th ed. Metals Park, OH, 1976.*

American Welding Society. *Welding Handbook,* Vol. 1, Miami, Fl, Latest Edition.

__________. *Welding Inspection.* Miami, FL, Latest Edition.

Anderson, R. C. *Visual Examination: Inspection of Metals,* Vol. 1. Metals Park, OH: American Society for Metals, 1983.*

Cary, H. B. *Modern Welding Technology.* Englewood Cliffs, NJ: Prentice-Hall, Inc., 1979.

Hobart Welding Guide. Troy, OH: Hobart School of Welding Technology, 1980.

McMaster, Robert C., ed. *Nondestructive Testing Handbook,* 1st ed. Columbus, OH: American Society for Nondestructive Testing, 1959.*

Megaw, E. D. "Factors Affecting Visual Inspection Accuracy," *Applied Ergonomics,* Cleveland, OH: IPC Business Press, March 1979.

Schoonard, J. W., et al. "Studies of Visual Inspection," *Ergonomics,* Vol. 16, No. 4. Philadelphia, PA: Taylor & Francis, Inc., 1973.

The Tools and Rules of Precision Measuring. Athol, MA: L. S. Starret Co., 1982.

Visual Examination Technology—101, 102, and 103. Charlotte, NC: EPRI NDE Center, 1983.

Welding and Fabrication Data Book. Cleveland, OH: Welding Design and Fabrication, 1984.

Yonemura, G. T. "Considerations and Standards for Visual Inspection Techniques," *Nondestructive Testing Standards—A Review.* Philadelphia, PA: American Society for Testing and Materials, 1977.

*Available from the American Society for Nondestructive Testing, Columbus, OH.

Appendix

Example Questions

The purpose of this appendix is to provide a guideline for the preparation of the General, Level I, and Level II written examinations. Extensive examples of representative questions for degree of difficulty, type, etc. are provided in separate question booklets, which can be obtained from ASNT Headquarters. These questions are intended as examples only and should not be used verbatim for qualification examinations.

Note: All questions and answers should be referenced to a recognized source.

Radiographic Testing Method

Level I

1. The most widely used unit of measurement for measuring the rate at which the output of a gamma-ray source decays is the
   a. curie
   b. roentgen
   c. half-life
   d. MeV

2. If an exposure time of 60 seconds were necessary using a 4 ft source-to-film distance for a particular exposure, what time would be necessary if a 2 ft source-to-film distance is used and all other variables remain the same?
   a. 120 seconds
   b. 30 seconds
   c. 15 seconds
   d. 240 seconds

3. The sharpness of the outline in the image of the radiograph is a measure of
   a. subject contrast
   b. radiographic definition
   c. radiographic contrast
   d. film contrast

Level II

1. When radiographing to the 2-2T quality level, an ASTM penetrameter for 2.5 in. steel has a thickness of
   a. 0.5 in.
   b. 2.5 mils
   c. 5 mils
   d. 50 mils

2. The approximate radiographic equivalence factors for steel and copper at 220 kV are 1.0 and 1.4, respectively. If it is desirable to radiograph a 0.5 in. plate of copper, what thickness of steel would require about the same exposure characteristics?
   a. 0.7 in. of steel
   b. 0.35 in. of steel
   c. 1.4 in. of steel
   d. 1.0 in. of steel

3. If a specimen were radiographed at 40 kV and again at 50 kV, with time compensation to give the radiographs the same density, which of the following statements would be true?
   a. The 40 kV exposure would have lower contrast and greater latitude than the 50 kV exposure.
   b. The 40 kV exposure would have higher contrast and greater latitude than the 50 kV exposure.
   c. The 50 kV exposure would have lower contrast and greater latitude than the 40 kV exposure.
   d. The 50 kV exposure would have higher contrast and greater latitude than the 40 kV exposure.

Magnetic Particle Testing Method

Level I

1. Which type of current has a "skin effect"?
   a. AC
   b. DC
   c. half-wave rectified
   d. full-wave rectified

2. The best type of magnetic field to use to inspect a tubular product for surface defects along its length is a
   a. longitudinal field
   b. circular field
   c. swinging field
   d. yoke magnetization

3. Which of the following is most often used for dry magnetic particle inspection?
   a. full-cycle DC
   b. half-wave AC
   c. high-voltage, low-amperage current
   d. DC from electrolytic cells

Level II

1. When testing a bar with an L/D ratio of four in a ten-turn coil, the required current would be
   a. 45,000 A
   b. unknown; more information is needed
   c. 18,000 A
   d. 1,125 A

2. Which of these cracks may appear as an irregular, checked, or scattered pattern of fine lines usually caused by local overheating?
   a. fatigue cracks
   b. grinding cracks
   c. crater cracks
   d. HAZ cracks

3. If a copper conductor is placed through a ferrous cylinder and a current is passed through the conductor, then the magnetic field (flux density) in the cylinder will be
   a. the same intensity and pattern as in the conductor
   b. greater than in the conductor
   c. less than in the conductor
   d. the same regardless of its proximity to the cylinder wall

Ultrasonic Testing Method

Level I

1. The amount of beam divergence from a crystal is primarily dependent on the
    a. type of test
    b. tightness of the crystal backing in the search unit
    c. frequency and crystal size
    d. refraction

2. On the area-amplitude ultrasonic-standard test blocks, the flat-bottomed holes in the blocks are
    a. all of the same diameter
    b. different in diameter, increasing by 1/64 in. increments from the No. 1 block to the No. 8 block
    c. largest in the No. 1 block and smallest in the No. 8 block
    d. drilled to different depths from the front surface of the test block

3. On many ultrasonic testing instruments, an operator conducting an immersion test can remove that portion of the screen presentation that represents water distance by adjusting a
    a. pulse-length control
    b. reject control
    c. sweep-delay control
    d. sweep-length control

Level II

1. If a contact angle-beam transducer produces a 45 degree shear wave in steel (V_s = 0.323 cm/s), the angle produced by the same transducer in an aluminum specimen (V_s = 0.310 cm/s) would be
    a. less than 45 degrees
    b. greater than 45 degrees
    c. 45 degrees
    d. more information is required

2. A discontinuity is located having an orientation such that its long axis is parallel to the sound beam. The indication from such a discontinuity will be
    a. large in proportion to the length of the discontinuity
    b. small in proportion to the length of the discontinuity
    c. representative of the length of the discontinuity
    d. such that complete loss of back-reflection will result

3. An ultrasonic longitudinal wave travels in aluminum with a velocity of 635,000 cm/s and has a frequency of 1 MHz. The wavelength of this ultrasonic wave is
    a. 6.35 ft
    b. 3.10 in
    c. 6.35 mm
    d. 30,000 Å

Liquid Penetrant Testing Method

Level I

1. Which of the following is generally the more acceptable method for cleaning parts prior to penetrant testing?
   a. sand-blasting
   b. wire-brushing
   c. grinding
   d. vapor-degreasing

2. The term used to define the tendency of certain liquids to penetrate into small openings such as cracks or fissures is
   a. saturation
   b. capillary action
   c. blotting
   d. wetting agent

3. Which of the following is the most commonly used method for removing non-water-washable visible dye penetrant from the surface of a test specimen?
   a. dipping in a solvent
   b. spraying
   c. hand-wiping
   d. blowing

Level II

1. When conducting a penetrant test, spherical indications on the surface of a part could be indicative of
   a. fatigue cracks
   b. porosity
   c. weld laps
   d. hot tears

2. A commonly used method of checking on the overall performance of a penetrant material system is by
   a. determining the viscosity of the penetrant
   b. measuring the wettability of the penetrant
   c. comparing two sections of artificially cracked specimens
   d. all of the above

3. Which of the following is a discontinuity that might be found in a forging?
   a. shrinkage crack
   b. lap
   c. hot tear
   d. lamination

Electromagnetic Testing Method

Eddy Current Testing Method

Level I

1. The impedance of an eddy current test coil will increase if the
   a. test frequency increases
   b. inductive reactance of the coil decreases
   c. inductance of the coil decreases
   d. resistance of the coil decreases

2. Which of the following test frequencies would produce eddy currents with the largest depth of penetration?
   a. 100 Hz
   b. 10 kHz
   c. 1 MHz
   d. 10 MHz

3. To generate measurable eddy currents in a test specimen, the specimen must be
   a. a conductor
   b. an insulator
   c. either a conductor or an insulator
   d. a ferromagnetic material

Level II

1. The fill factor when a 0.5 in. diameter bar is inserted in a 1 in. diameter coil is
   a. 0.5 (50 percent)
   b. 0.75 (75 percent)
   c. 1.0 (100 percent)
   d. 0.25 (25 percent)

2. If the characteristic frequency (f_g) of a material is 125 Hz, the test frequency required to give an f/f_g ratio of 10 would be
   a. 1.25 Hz
   b. 12.5 Hz
   c. 1.25 kHz
   d. 12.5 kHz

3. For age-hardened aluminum and titanium alloys, changes in hardness are indicated by changes in
   a. retentivity
   b. permeability
   c. conductivity
   d. magnetostriction

Flux Leakage Testing Method

Level I

1. Flux leakage inspection can normally be applied to
   a. ferromagnetic and nonmagnetic material
   b. nonmagnetic materials only
   c. ferromagnetic materials only
   d. nonconductors only

2. The ratio B/H is equivalent to a material's
   a. field strength
   b. reluctance
   c. permittivity
   d. permeability
   e. relative permeability

3. In the flux leakage examination of tubular products, which of the following discontinuities can be detected?
   a. longitudinally oriented
   b. transversely oriented
   c. slivers
   d. all of the above

Level II

1. The highest sensitivity of a Hall generator is obtained when the direction of the magnetic field in relation to the largest surface of the Hall probe is
   a. parallel
   b. at an angle of 45 degrees
   c. at an angle of 30 or 60 degrees
   d. perpendicular
   e. none of the above

2. What particular type of defect is **not** indicated by flux leakage techniques?
   a. overlap
   b. grain-boundary crack
   c. slag inclusion with crack
   d. surface contamination
   e. longitudinal seam

3. In the examination of tubular products where the flux sensor measures the leakage field at the outside surface of the tube,
   a. outside-diameter discontinuities are detected.
   b. both outside-diameter and inside-diameter discontinuities may be detected.
   c. both outside-diameter and inside-diameter discontinuities can be detected but generally cannot be distinguished from each other.
   d. both outside-diameter and inside-diameter discontinuities can be detected and generally can be distinguished from each other.

Neutron Radiographic Testing Method

Level I

1. Neutron penetration is greatest in which of the following materials?
    a. hydrogenous material
    b. water
    c. lead
    d. boron carbide

2. Gadolinium conversion screens are usually mounted in rigid holders called
    a. film racks
    b. cassettes
    c. emulsifiers
    d. diaphragms

3. Which element is commonly used for direct neutron radiography?
    a. Cd
    b. In
    c. Dy
    d. Gd

Level II

1. Which of the following conversion screens has the longest half-life?
    a. dysprosium
    b. indium
    c. cadmium
    d. gadolinium

2. Neutron radiography can be used for inspecting which of the following applications?
    a. presence of explosives in a metal device
    b. presence of foreign materials such as oil
    c. lubricants in metal systems
    d. hydrogen content in metals
    e. all of the above

3. Real-time imaging of thermal neutron radiography can be performed with which of the following detectors?
    a. gadolinium
    b. dysprosium
    c. zinc sulfide + lithium fluoride
    d. europium

Leak Testing Method

Bubble Leak Testing Method

Level I

1. Before performing a vacuum box leak test, which of the following should be checked for required calibration?
   a. leak-detector solution
   b. evacuation device or equipment
   c. lighting equipment
   d. pressure (or vacuum) gage

2. Which factor can most affect the sensitivity attainable by a pressure bubble leak test?
   a. operator alertness and technique
   b. size and shape of the test specimen
   c. time of day testing is performed
   d. number of test technicians

3. The letters "psia" mean
   a. pressure referred to National Institute of Standards and Technology's absolute pressure
   b. pascals per square inch absolute
   c. pressure standard in absolute units
   d. pounds per square inch absolute

Level II

1. Which of the following directly determines the size of the bubble formation when testing using the bubble test method?
   a. method of application of bubble solution
   b. ambient temperature and barometric pressure
   c. amount of leakage from a defect or leak
   d. size of the test specimen

2. When a vacuum gage is marked with a range of 0–30 with the notation "vacuum" on the face, the units of measurement are
   a. inches of mercury
   b. pounds per square inch
   c. centimeters of vacuum
   d. feet of water

3. The type of leaks that are most likely to go undetected during a bubble leak test are
   a. very small leaks and very large leaks
   b. leaks occurring at welded joints
   c. corner-configuration joints
   d. all of the above

Halogen Diode Detector Leak Testing Method

Level I

1. Good operating practice dictates that the period of time to allow for warmup of the halogen diode detector prior to calibrating is
   a. 30 minutes
   b. 15 minutes
   c. one hour
   d. as recommended by the manufacturer

2. While adjusting a reservoir-type variable-halogen standard leak, the operator accidentally vents the gas from the only standard leak available. Which of the following actions would quickly resolve the problem?
    a. Replace the standard leak.
    b. Replace the cylinder in the standard leak.
    c. Recharge the standard leak.
    d. Send the standard leak to the manufacturer for recharging.

3. While performing a halogen-diode detector test, the leak detector becomes difficult to zero, and the pointer on the leak rate meter repeatedly swings up scale. The most likely cause of the problem could be the use of too high a sensitivity range, a shorted element, an excessive heater voltage, or
    a. poor airflow
    b. a sensing element that is too new
    c. a high halogen background
    d. a faulty leak-rate meter

Level II

1. Most leaks detected during a halogen sniffer test could have been detected and usually can be verified by
    a. a bubble leak test
    b. an ultrasonic examination
    c. a visual examination
    d. a pressure change test

2. The presence of small traces of halogen vapors in the halogen diode detector
    a. increases the emission of negative ions
    d. decreases the emission of positive ions
    c. increases the emission of positive ions
    d. decreases the emission of negative ions

3. A halogen standard leak of a certain size produces a known signal on a halogen leak detector. To receive this same intensity signal on the instrument during the test of an object containing a 2 percent by volume halogen-air mixture, the size of the leak in the object causing the signal would theoretically have to be at least __________ times larger than the standard leak.
    a. 20
    b. 50
    c. 40
    d. 10

Mass Spectrometer Leak Testing Method

Level I

1. The sensitivity of a mass spectrometer leak-detection system is the mass flow rate of tracer gas
    a. that gives a maximum measurable signal
    b. that gives a minimum measurable signal
    c. at standard temperature and pressure
    d. in a leak

2. The diffusion pump and mechanical forepump in a mass spectrometer leak-detection system
    a. use the same type of oil
    b. use different types of oil
    c. operate using the same motor
    d. use the same principle of operation

3. The helium mass spectrometer detector-probe pressure-test technique is
    a. a quantitative test
    b. a qualitative test
    c. a semiautomatic test
    d. none of the above

1. A torr is defined as
    a. 14.7 psia
    b. 1 μm of Hg
    c. 1/760 of a standard atmosphere
    d. 760 mm of Hg

2. When conducting a helium mass-spectrometer test of a vacuum vessel in the pressure range of 10^{-4} to 10^{-8} mm Hg, which type gage could be used to measure the pressure?
    a. alphatron gage
    b. thermionic ionization gage
    c. Pirani gage
    d. thermocouple gage

3. Helium standard leaks in the range of 10^{-6} to 10^{-10} atm. cc/s are known in general terms as
    a. reservoir standard leaks
    b. capillary standard leaks
    c. permeation standard leaks
    d. adjustable standard leaks

Pressure Change Measurement Leak Testing Method

Level I

1. A pressure of 66.0 psig, in terms of absolute pressure at sea level and standard temperature, would be approximately
    a. 96.0 psia
    b. 80.7 psia
    c. 51.3 psia
    d. 36.0 psia

2. When conducting a long-duration pressure change test, it is necessary to measure either absolute pressure or gage pressure plus barometric pressure because the barometric pressure will
    a. always fall
    b. always rise
    c. remain constant
    d. tend to vary

3. Which one of the following is the correct relationship for converting temperature in degrees Rankine (°R) to temperature in degrees Kelvin (°K)?
    a. °K = (5/9)°R
    b. °K = (5/9)°R + 273
    c. °K = 460 + °R
    d. °K = 273°R

Level II

1. When a system's internal dry bulb's internal temperature and, in turn, total pressure, increase during a pressure change leakage-rate test, the water vapor pressure in the system under test would normally
    a. increase
    b. remain the same
    c. decrease
    d. oscillate

2. For a pneumatically pressurized constant-volume system at an internal temperature of 27°C, what approximate percentage change in the system absolute pressure can be expected for a system internal temperature change of 1°C?
    a. 3 percent
    b. 6 percent
    c. 0.3 percent
    d. 10 percent

3. One set of internal dry bulb temperature data for a pressure change leakage rate test is

$$(T_1 + T_2 + T_3)/3 = 71.87°F$$

$$(T_4 + T_5)/2 = 72.32°F$$

$$(T_6 + T_7)/2 = 72.68°F$$

$$(T_8 + T_9 + T_{10})/3 = 73.07°F$$

For each of these four sections of this system, the respective weighting factors are 0.27, 0.18, 0.22, and 0.33. The mean absolute dry bulb temperature of system air for this test data point is
    a. 532.53°R
    b. 345.53°K
    c. 532.48°R
    d. 532.48°K

Acoustic Emission Testing Method

Level I

1. A qualitative description of the sustained signal level produced by rapidly occurring acoustic emission events is the accepted definition for
   a. burst emission
   b. acoustic emission signature
   c. acoustic emission signal
   d. continuous emission
   e. none of the above

2. Attenuation of a wave is best defined by which statement?
   a. a decrease in frequency with distance traveled
   b. a decrease in amplitude with distance traveled
   c. a decrease in wave speed with distance traveled
   d. a change in direction as a function of time

3. The number of times the acoustic emission signal exceeds a preset threshold during any selected portion of a test is called the
   a. acoustic emission response
   b. acoustic emission count
   c. acoustic emission count rate
   d. acoustic emission energy
   e. none of the above

Level II

1. When detecting impulsive acoustic emission signals on large objects, the peak of the signals normally decreases with increasing distance from the source. This alteration, dependent on distance, must be explained by
   a. absorption: i.e., the elastic pulse gradually converts into heat
   b. dispersion: i.e., the pulse gradually spreads out in time because the different waves involved travel with different velocities
   c. the geometric factors: i.e., the energy in the pulse is distributed into ever-larger volumes
   d. all of the above

2. Which of the following factors will tend to produce *low-amplitude* acoustic emission response during a tensile test?
   a. low temperature
   b. high strain rate
   c. plastic deformation
   d. crack propagation

3. The Kaiser effect is
   a. valid only when testing composites
   b. a physical law of nature that is never violated
   c. not applicable when an rms recording is being made
   d. the absence of detectable acoustic emission until previously applied stress levels are exceeded

Visual Testing Method

Level I

1. Which of the following is true?
   a. All discontinuities are defects.
   b. Defects that affect the product's usefulness are called discontinuities.
   c. Discontinuities that affect the product's usefulness are called defects.
   d. All discontinuities are unacceptable.

2. The dimension indicated on the sketch of a micrometer is
   a. 0.128 in.
   b. 0.235 in.
   c. 0.126 in.
   d. 0.328 in.

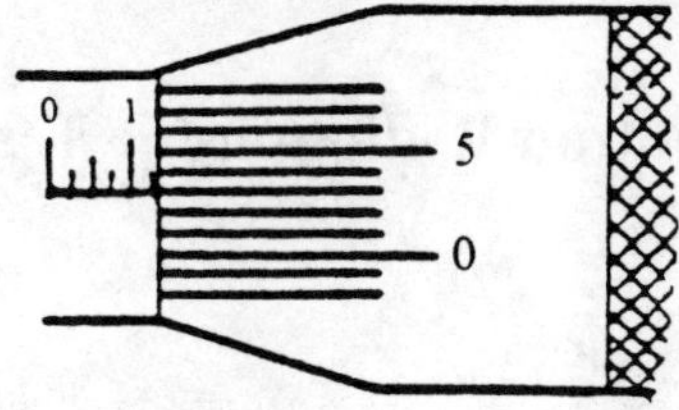

3. As a visual examiner, you shall have your eyes checked at least
   a. every 3 months
   b. every 6 months
   c. every year
   d. every 3 years

Level II

1. Handheld magnifiers should fall into which of the following ranges?
   a. 2X to 4X
   b. 5X to 10X
   c. 10X to 20X
   d. 20X to 30X

2. Visual examiners who perform visual exams using borescopes and fiberscopes must be
   a. color-blind
   b. able to meet far-vision requirements (Snellen 20/30)
   c. competent in their use
   d. ambidextrous

3. A narrow field of view produces
   a. higher magnification and a greater depth of field
   b. higher magnification and a shorter depth of field
   c. less magnification and a greater depth of field
   d. less magnification and a shorter depth of field

Answers to Example Questions

Radiographic Testing Method

Level I
1. a
2. c
3. b

Level II
1. d
2. a
3. c

Magnetic Particle Testing Method

Level I
1. a
2. b
3. b

Level II
1. d
2. b
3. b

Ultrasonic Testing Method

Level I
1. c
2. b
3. c

Level II
1. a
2. b
3. c

Liquid Penetrant Testing Method

Level I
1. d
2. b
3. c

Level II
1. b
2. c
3. b

Electromagnetic Testing Method
Eddy Current Testing Method
Level I
1. a
2. a
3. a

Level II
1. d
2. c
3. c

Flux Leakage Testing Method
Level I
1. c
2. d
3. d

Level II
1. d
2. d
3. d

Neutron Radiographic Testing Method

Level I
1. c
2. b
3. d

Level II
1. a
2. e
3. c

Leak Testing Method
Bubble Leak Testing Method
Level I
1. d
2. a
3. d

Level II
1. c
2. a
3. a

Halogen Diode Detector Leak Testing Method
Level I
1. d
2. c
3. c

Level II
1. a
2. c
3. b

Mass Spectrometer Leak Testing Method
Level I
1. b
2. b
3. b

Level II
1. c
2. b
3. c

Pressure Change Measurement
Leak Testing Method
Level I
1. b
2. d
3. a

Level II
1. a
2. c
3. a

Acoustic Emission Testing Method

Level I
1. d
2. b
3. b

Level II
1. d
2. c
3. d

Visual Testing Method

Level I
1. c
2. a
3. c

Level II
1. b
2. c
3. b